Thermal Power Plants - Advanced Applications

Thermal Power Plants - Advanced Applications

Editor

Azat Solak

scitus
academics

Thermal Power Plants - Advanced Applications
Edited by **Azat Solak**

ISBN: 978-1-68117-616-1
Library of Congress Control Number: 2016946474

616, Corporate Way, Suite 2, 4766,
Valley Cottage, NY 10989

www.scitusacademics.com

Notice

Reasonable efforts have been made to publish reliable data and views articulated in the chapters are those of the individual contributors, and not necessarily those of the editors or publishers. Editors or publishers are not responsible for the accuracy of the information in the published chapters or consequences of their use. The publisher believes no responsibility for any damage or grievance to the persons or property arising out of the use of any materials, instructions, methods or thoughts in the book. The editors and the publisher have attempted to trace the copyright holders of all material reproduced in this publication and apologize to copyright holders if permission has not been obtained. If any copyright holder has not been acknowledged, please write to us so we may rectify.

Printed in United States of America on Acid Free Paper

Preface

Thermal power plants are among the most important process industries for professionals in engineering. In recent decades, the power industry has actually dealt with a variety of critical issues; however, the growing demand for power is met by the most important challenge in efficient and sustainable ways. Practitioners of power plant engineers not only take care of the maintenance and operation of the plant, but also take care of a variety of activities, including development and research, from power generation to the environment of power plants. A thermal power station is a power station in which steam is used to a steam turbine. This turbine is attached to a generator. The water is then condensed, and can be used again. This is called the Rankine cycle. You will find various methods that can be used to heat the water; many types of thermal power plants are available. The theory of a thermal power plant or the possible operation of a thermal power plant is so easy. A power generation plant consists mainly of a steam turbine alternator. Vapor is produced from high-pressure boilers.

Addressing different aspects of a thermal power plant, giving the subject a new dimension, focusing on operating practices and troubleshooting, technology and design, the book presents reliability-based tools used to define the performance of complex systems and introduces the basic concepts of reliability, maintenance and risk analysis to be used as power plant performance instruments. The book is valuable both for advanced graduates and researchers and, of course, for engineers working in power plants.

Table of Contents

Chapter 1

Application of System Analysis for Thermal Power Plant Heat Rate Improvement

M.N. Lakhoua[1], M. Harrabi[2] and M. Lakhoua[2]

[1] Université de Tunis El Manar, Ecole Nationale d'Ingénieurs de Tunis, LR11ES20 Analyse, Conception et Commande des Systèmes, Tunis, Tunisie

[2] Société Tunisienne de l'Electricité et du Gaz, Tunisie

1. INTRODUCTION

In order to improve the performance of a thermal power plant (TPP), it is necessarily to adopt performance monitoring and heat rate improvement. To improve efficiency, the engineer must knew the heat input, the mass of fuel, the fuel analysis and the kW rating generation in order to determine the actual heat rate. After the actual heat rate calculated and understood, losses must be identified and understood. Good communication and teamwork between the engineer and staff within the TPP is essential to success [1-2].

In fact, the heat rate is defined in units of Btu /kWh (KJ / kWh) and is simply the amount of heat input into a system divided by the amount of power generated by of a system [1].

The calculation of the heat rate enable to inform us on the state of the TPP and help the engineer to take out the reasons of the degradation of the TPP heat rate in order to reach the better one recorded at the time of acceptance test when the equipment was new and the TPP was operated at optimum. Therefore, this TPP heat rate value is realistic and attainable for it has been achieved before [1-2].

The global efficiency of a TPP is tributary of a certain number of factors and mainly of the furnace efficiency. Otherwise, there is place to also notice that with regards to the turbine and the alternator that are facilities of big importance in the constitution of a TPP, the degradation of their respective efficiencies hardly takes place long-term of a manner appreciable and this by reason of the ageing of some of their organs as: stationary and mobile aubages usury; increase of the internal flights; usury of alternator insulations, etc.

The objective of this paper is to analyze the different losses of a TPP therefore to implant solutions in order to act in time and to improve its efficiency. Appropriate performance parameters can enable the performance engineer to either immediately correct performance or estimate when it would be cost effective to make corrections. In fact, the performance parameters measure how well a TPP produces electricity.

These actions or decisions are [1]:

- Improve TPP operation;
- Predictive maintenance;
- Comparison of actual to expected performance;
- Improved economic dispatch of TPP;
- Reduce uncertainty in actual costs for better MW sales.

This paper can be loosely divided into five parts. First, we present the functionality of TPP. Second, we present the boiler and steam turbine efficiency calculations. In section 3, we present the methodology of the analysis based on the Objectives Oriented Project Planning (OOPP) method. In section 4, we present the results of the application of system analysis for determining the possible losses for the degradation of the TPP heat rate. The last section presents a conclusion about the advantages and inconveniences of the analysis presented of the TPP heat rate improvement.

2. FUNCTIONALITY OF A THERMAL POWER PLANT

Thermal power plant (TPP) is a power plant in which the prime mover is steam driven. Water is heated, turns into steam and spins a steam turbine which drives an electrical generator. After it passes through the turbine, the steam is condensed in a condenser. The greatest variation in the design of TPPs is due to the different fuel sources. Some prefer to use the term energy center because such facilities convert forms of heat energy into electrical energy [3-5].

In TPPs, mechanical power is produced by a heat engine which transforms thermal energy, often from combustion of a fuel, into rotational energy. Most TPPs produce steam, and these are sometimes called steam power plants. TPPs are classified by the type of fuel and the type of prime mover installed (Figure 1).

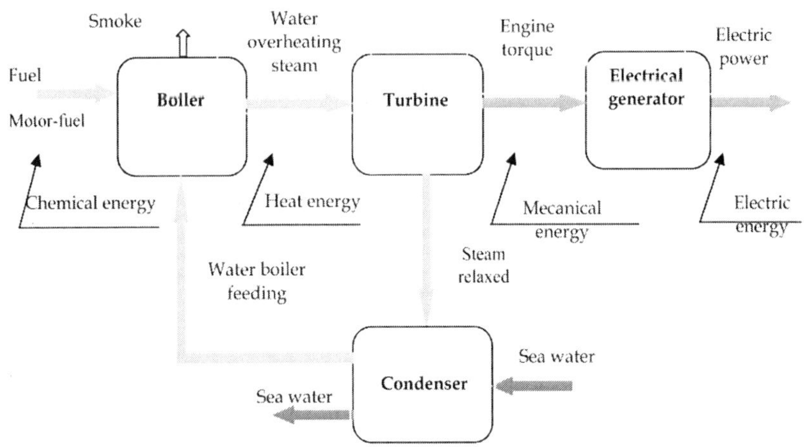

Figure 1. Functionality of a TPP.

The electric efficiency of a conventional TPP, considered as saleable energy produced at the plant busbars compared with the heating value of the fuel consumed, is typically 33 to 48% efficient, limited as all heat engines are by the laws of thermodynamics. The rest of the energy must leave the plant in the form of heat.

Since the efficiency of the plant is fundamentally limited by the ratio of the absolute temperatures of the steam at turbine input and output, efficiency improvements require use of higher temperature, and therefore higher pressure, steam.

This overheated steam drags the HP rotor (high pressure) of the turbine in rotation and relaxes to the exit of the HP body of the turbine, so it comes back again in the furnace to be until 540° after, it will be sent back to the MP body (intermediate pressure) then to the BP body (low pressure) of the turbine.

During these steps, the calorific energy is transformed in available mechanical energy on the turbine. Thus, this mechanical energy will be transmitted to the alternator, being a generator of alternating current, in the goal to produce the electric energy.

After the condensation, water will be transmitted thanks to pumps of extraction in the station of BP to be warmed progressively before being sent back to the furnace through the intermediary of the food pumps.

This warms progressive of water has for goal to increase the output of the furnace and to avoid all thermal constraints on its partitions. And this station of water is composed of a certain number of intersections that is nourished in steam of the three bodies of the turbine. Finally, the cycle reproduces indefinitely since steam and water circulate in a closed circuit.

During this cycle water recovers the calorific energy in the boiler that it restores at the time of its detente in the turbine as a mechanical energy to the rotor of the turbine. The rotor of the turbine being harnessed to the rotor excited of the alternator, the mechanical energy of the turbine is transformed then in electric energy in the alternator.

Turbine constitutes an evolution exploiting principal's advantages of turbo machines: mass power and elevated volume power; improved efficiency by the multiplication of detente floors [6-8].

Indeed, a steam turbine is a thermal motor with external combustion, functioning according to the thermodynamic cycle Clausius-Rankine. This cycle is distinguished by the state change affecting the motor fluid that is the water steam (Figure 2).

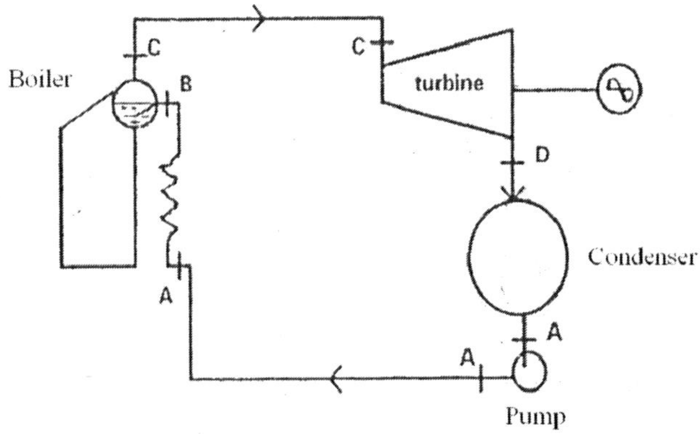

Figure 2. Clausius-Rankine cycle.

The efficiency grows with the steam pressure and with the overheat temperature. However, the increase of these features is limited by the water content in steam in the end of detente.

Indeed, the detente curve can reach the saturation curve with formation of droplets that is harmful to the efficiency of the last floors of detente. The content in liquid water of the mixture must be limited to 15 or 20%. At the end, it is the condenser pressure that fixes the admissible limits of pressure and temperature.

2.1. Boiler Accessories

The boiler is a steam generator that assures the spraying of water (Figure 3). At this level operates the transformation of the chemical energy in calorific energy by combustion of a mixture "air-fuel".

Figure 3 presents an example of a boiler in a TPP in Tunisia.

The boiler is composed by different elements:

- The combustion room

It constitutes a surrounding wall of contiguous tubular bundles inside in the water circulates. It is in this combustion room the transformation of the chemical energy in calorific energy by combustion of a mixture "air-fuel". This calorific energy frees a quantity of heat that will be transmitted to water to produce the steam of water in a temperature and under a very determined pressure.

Figure 3. Example of a boiler of a TPP.

- The economiser

It has for role to recover a part of calories remaining in the gases of combustion to increase the temperature of the feeding water what will have for effect the increase of the thermal output of the installation and the elevated thermal constraint suppression in the metal of the reservoir.

- The ball of the boiler

To the exit of the economizer, the water of feeding goes up toward a reservoir situated in the part superior of the boiler called ball of the boiler that constitutes a surrounding wall in sheet metal in which is the liquid phase and the phase steam of the feeding water.

- Superheater / Reheater

It is an intersection of heat constituted of tubular bundles re-serving the gases of combustion directly; therefore submissive to the most elevated temperatures of the combustion room. Steam coming from the ball is humid; it passes therefore in tubes of the heater where its temperature is raised to relatively constant pressure.

After having undergone a first detente in the high pressure body of the turbine, steam comes back to the generator of steam and enter in an intersection called primary reheater of temperature 330°C then it crosses the final re-heater of temperature 540°C, then it is sent toward the intermediate pressure body of the turbine. To the exit of the intermediate pressure body steam passes in the body low pressure of the turbine.

- Burners

The burner is the most important component to light the natural gas fuel-oil. Le role of the burner is of creates a zone of ignition to the sufficient temperature to maintain the combustion and to provide the necessary air mixture. The boiler include seven floors each one is equipped of four burners.

2.2. Steam Turbine

A turbine is constituted of a rotor composed of a tree on which is fixed the dawns and a stator of composed of a structural cover of the stationary deflectors, generally organized of two parts according to an axial plan [9-12].

The turbine is composed of a segmented admission tore and a controlled exhaust divergent toward the condenser. The stationary deflector function is to assure all or one of the detente while forming a nozzles network and to modify the direction of the out-flow retiring of the previous floor.

A steam turbine is composed of one or several floors assuring each two functions:

- The steam detente that corresponds to the conversion of the potential energy in kinetic energy;

- The conversion of the kinetic energy in rotation couple of the machine by the mobile aubages.

The steam turbines are often classified in two big categories combined in the same machine:

- Turbines to action in which the detente makes himself solely in the stationary aubages. They are well adapted to strong pressure floors and are better suitable to the debit regulation. Their construction is more expensive and their use for the first floors.

- The jet-propelled turbines in which the detente is distributed between stationary and mobile aubages. The degree of reaction is defined by the distribution of the detente between aubages. They are better suitable to bass pressure floors and their cost is weaker.

The realization of turbines requires the recourse to greatly allied steels (Cr-Ni-Va) to resist the thermal, mechanical constraints (centrifugal force) and chemical (steam corrosion). The first two constraints limit the diameter and therefore the capable debit of the last floors. So dawns besides of one meter of length already put serious problems of realization. Besides, the radial heterogeneity of speeds imposes a variable impact of the dawn that present then a left shape whose machining is complex [9-12].

The principal favour of steam turbines is to be external combustion motors. Of this fact, all fuels (gas, fuel-oil, coal, vestigial, geothermal heat) can be supplied it with steam. The efficiency can reach some elevated values and reduced working expenses (specific consumption of 2300 kcal/kWh and 3400 kcal/kWh for gas turbines). The cost and the complexity of facilities are the most often reserved to the elevated power facilities. But in particular cases, motors and gas turbines are better adapted below about 10 MW.

Figure 4 presents an example of steam turbines in a TPP in Tunisia.

3. BOILER AND STEAM TURBINE EFFICIENCY CALCULATIONS

In this paragraph, we present some methods used in order to determine the boiler efficiency and the steam turbine efficiency [13-15].

Figure 4. **Disposition of steam turbines in a TPP.**

3.1. Boiler Efficiency Calculations

Among the multiple factors that can degrade the boiler efficiency, there is place to mention what follows as an example:

- bad combustion following a bad working of the regulation;
- bad quality of the fuel used;
- heating inadequate of the used fuel;
- flights of water and steam;
- flights of air (comburant);
- encrassement of the boiler;
- encrassement of the air heating device...

Boiler efficiency can be calculated by one of two methods: the Input-Output method or the method of heat losses [1].

3.1.1. Input / Output Method

The expression of the boiler efficiency of the TPP is given by:

$$R_{ch} = \frac{\text{Output}}{\text{Input}}. \tag{1}$$

R_{ch}: Boiler efficiency.

The exit is defined by the sum of heats absorbed by the used fluid (water-steam).

The entrance is defined by the total energy introduced in the boiler.

The boiler efficiency is given by the following expression:

$$R_{ch} = \frac{[Q_{eal}(H_{sh}-H_{al})+Q_{injsh}(H_{sh}-H_{injsh})+Q_{vrh}(H_{rh}-H_{erh})+Q_{injrh}(H_{erh}-H_{injrh})]}{F_{h1}.Qf+B_{e}}. \tag{2}$$

- Q_{eal} : debit water (kg/h);
- Q_{injsh}: debit water injection;
- Q_{injrh}: debit of the water of steam (Kg/h).
- H_{sh}: enthalpy of steam to the exit;
- H_{al}: enthalpy of water in entrance economizer;
- H_{rh}: enthalpy of water (kcal/kg).
- H_{erh}: enthalpy of water exit HP (kcal/kg).
- H_{injrh}: enthalpy of water (kcal/kg).
- F_{h1}: superior calorific power of fuel used (kcal/kg);

- Q_f : debit fued (t/h);
- B_e : total of heat introduced.

3.1.2. Method Of Heat Losses

The boiler efficiency is given by the following expression:

$$R_{ch} = \frac{F_h - C_{per}}{F_h} \quad (3)$$

F_h : Total energy introduced in the boiler.

C_{per}: Sum of the calorific losses at the level of the boiler.

In the case of fuel, the total heat introduced in the boiler comes from fuel, of the air of combustion and the steam of atomization.

The calorific losses that one meets in a boiler are essentially owed to the heat carried away by the gases of combustion, to the presence of water in fuel as well as the existing humidity in the air of combustion.

3.2. Steam Turbine Efficiency Calculation

The calculation of the efficiency of bodies of the turbine with the difference enthalpy method is very useful for the assessment of the cleaning degree of the steam course in the body of the turbine.

The efficiency of the turbine is defined as the report between the real difference enthalpy (DHIHP) and the isotropy difference enthalpy (DHIHP) of steam crossing the HP body.

For the BP body this method is not applicable because of the title of steam to the BP exit (humid steam).

The efficiency of the HP body (ηHP) is defined as follows [1]:

$$\eta_{HP} = \frac{D_{HRHP}}{D_{HIHP}} \times 100 \quad (4)$$

The real enthalpy difference in the HP body is given by:

$$D_{HRHP} = H_{vap} - H_{erh} \left(\text{kcal/kg}\right) \quad (5)$$

H_{vap}: enthalpy of steam overheated admission HP turbine.

H_{erh}: enthalpy of steam to overheat HP exit.

The difference isotropy enthalpy in the HP body is given by:

$$D_{HIHP} = H_{vap} - H_{ith} \quad (6)$$

H_{ith}: final enthalpy of steam overheated for an expansion isotropy (S= constant) of the admission until exit of the HP turbine.

In addition the indicated thermodynamic losses appear in the machine external energy losses provoked mainly by rubbings mechanical landings furniture and flights. The efficiency of the turbine must take into account these losses.

The efficiency of the turbine is:

$$R_t = R_{th} \times R_{vol} \times R_{mec} \quad (7)$$

While the volumetric efficiency (R_{vol}) is equal to:

$$Rvol = 1 - \frac{g}{G} \quad (8)$$

g: debit of flight ; G: debit weight.

4. METHODOLOGY OF ANALYSIS

There are many methods that have been used to enhance participation in Information System (IS) planning and requirements analysis. We review some methods here because we think them to be fairly representative of the general kinds of methods in use. The methods include Delphi, focus groups, SADT (Structured Analysis Design Technique), multiple criteria decision-making (MCDM), total quality management (TQM) and OOPP method (Objectives Oriented Project Planning).

The objective of the Delphi method is to acquire and aggregate knowledge from multiple experts so that participants can find a consensus solution to a problem [16].

A second distinct method is focus groups (or focused group interviews). This method relies on team or group dynamics to generate as many ideas as possible. Focus groups been used for decades by marketing researchers to understand customer product preferences [17].

MCDM views requirements gathering and analysis as a problem requiring individual interviews [18]. Analysts using MCDM focus primarily on analysis of the collected data to reveal users' requirements, rather than on resolving or negotiating ambiguities. The objective is to find an optimal solution for the problem of conflicting values and objectives, where the problem is modelled as a set of quantitative values requiring optimization.

TQM is a way to include the customer in development process, to improve product quality. In a TQM project, data gathering for customers needs, i.e., requirements elicitation may be done with QFD [19].

The SADT method represent attempts to apply the concept of focus groups specifically to information systems planning, eliciting data from groups of stakeholders or organizational teams [20]. They are characterized by their use of predetermined roles for group/team members and the use of graphically structured diagrams. SADT enables capturing of a proposed system's functions and data flows among the functions.

The OOPP method, used in this survey, is considered like a tool of communication, analysis and scheduling of project, whatever is its nature, its situation, its complexity and its sensitivity [21-24].

In this part, we present the OOPP method that we use in order to determine the different losses of the TPP.

4.1. Oopp Method

This method is used more and more by several financial backers (World Bank, Union European, bilateral Cooperation...). It is also used to take to terms of development projects, of cooperation (Germany, Canada, Belgium...) or other. It gave a good satisfaction at the time of its exploitation and several researches have been done very well to develop tools and to prove its strength for the scheduling of projects.

The descriptive documentation of the OOPP method, indicate that the logic of the OOPP method is not in principle limited not to a type of a determined problematic. Nevertheless, in practice the method is more appropriated to the following interventions: projects of the technical cooperation and projects of investments with economic and / or social objective.

The OOPP method which is also referred to Logical Framework Approach (LFA) is a structured meeting process. This approach is based on four essential steps: Problem Analysis, Objectives Analysis, Alternatives Analysis and Activities Planning. It seeks to identify the major current problems using cause-effect analysis and search for the best strategy to alleviate these identified problems [21-24].

The first step of "Problem Analysis" seeks to get consensus on the detailed aspects of the problem. The first procedure in problem analysis is brainstorming. All participants are invited to write their problem ideas on small cards. The participants may write as many cards as they wish. The participants group the cards or look for cause-effect relationship between the themes on the cards by arranging the cards to form a problem tree (Figure 5).

In the step of "Objectives Analysis" the problem statements are converted into objective statements and if possible into an objective tree (Figure 6). Just as the problem tree shows cause-effect relationships, the objective tree shows means-end relationships. The means-end relationships show the means by which the project can achieve the desired ends or future desirable conditions.

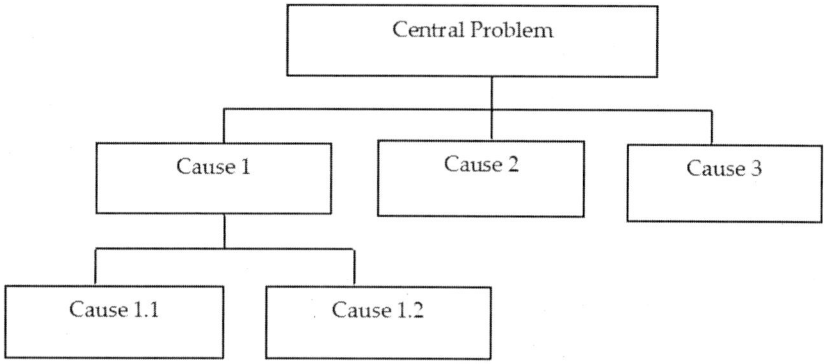

Figure 5. Problem tree of the OOPP method.

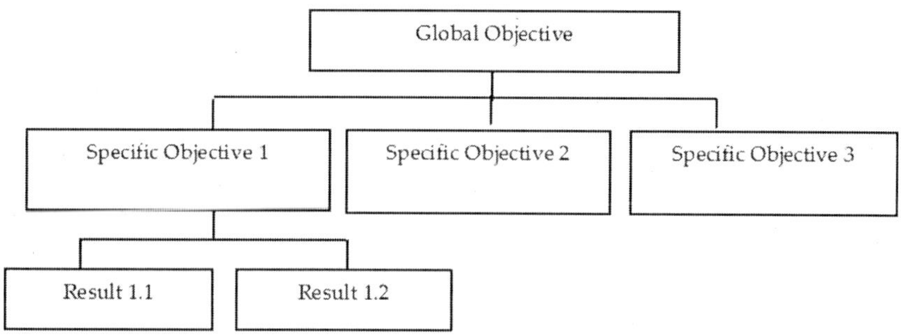

Figure 6. Objective tree of the OOPP method.

The objective tree usually shows the large number of possible strategies or means-end links that could contribute to a solution to the problem. Since there will be a limit to the resources that can be applied to the project, it is necessary for the participants to examine these alternatives and select the most promising strategy. This step is called "Alternatives Analysis".

After selection of the decision criteria, these are applied in order to select one or more means-end chains to become the set of objectives that will form the project strategy.

After defining the objectives and specifying how they will be measured (Objectively Verifiable Indicators: OVIs) and where and how that information will be found (Means of Verification: MOVs) we get to the detailed planning phase: "Activities Planning". We determine what activities are required to achieve each objective. It is tempting to say; always start at the situation analysis stage, and from there determine who are the stakeholders.

4.2. Some Researches On The Oopp Method

We present some studies of the OOPP method in IS planning that have been presented in various researches:

Researchers, Gu & al. [25] have presented an object-oriented approach to the development of a generative process planning system. The system consists of three functional modules: object-oriented product model module, object-oriented manufacturing facility model module, and object-oriented process planner.

Researcher, Hill [26] has question the appropriateness of highly structured strategic planning approaches in situations of complexity and change, using the Cambodian-German Health Project as a case study. He has demonstrated the limitations of these planning processes in complex situations of high uncertainty, with little reliable information and a rapidly changing environment.

Researchers, Peffers & al. [27] have used information theory to justify the use of a method to help managers better understand what new Information Technology applications and features will be most valued by users and why and apply this method in a case study involving the development of financial service applications for mobile devices.

Researchers, Killich & al. [28] have presented the experiences and results of the development and implementation of a software-tool for a SME-network in the German automotive supply chain industry. The tool called TeamUp enables the communication of experts as well as the coordination of discussion groups in order to make use of synergetic potentials.

4.3. Refining The Oopp Method

The application of the OOPP method to the identification of activities of work stations is important. The management of a system is conditioned notably by ties between its Entities Activities (EA), being able to be according to their hierarchical Specific Objective level (SO), of Results (R), of Activities (TO), of Under-activities (S), of Tasks (T)... These ties are materialized in fact by exchanges of information (If) produced by certain activities and consumed by others. The restraint of these ties requires an extension of the method. This new extension permits to identify the manner to execute these activities and to manage the different phases of the system.

An effort has been provided in order to refine the OOPP method. The OOPP method has been spread and a new denomination MISDIP (Method of Specification, Development and Implementation of Project) was adopted. The MISDIP method adopts the OOPP analysis and the complete it to specify the system of organization, to specify the system of information, and to contribute to its development and implementation [29].

We defined the Method of Informational Analysis by Objectives (MIAO) [30] permitting to elaborate an information matrix that permits to analyze the informational exchange process between activities.

In fact, the identification and the analysis of the information exchanged by the activities indicate the dynamics and the communication between the elements of the system that we propose to study or to manage. So, an information matrix was defined. This matrix establishes a correlation between activities and their information. The information concerning an activity can be classified in two categories:

- The imported information by an activity is supposed to be available: it is either produced by other activity of the system, or coming from outside;
- The produced information by an activity reflects the state of this activity. This last information may be exploited by other activities of the project.

In fact, the information produced by an activity can be considered like a transformation of imported information by this activity.

In order to specify this information, we define an information matrix (Table 1) associated to OOPP analysis enabling to determine the relations between the activities or between the concerned structures, to identify the information sources and to determine the manner in which the information is exploited.

Table 1. The information matrix of the MIAO.

N°	Code	Activity	If_1	If_2	If_3	If_4	If_5	If_6	If_7	If_8	If_n
1		A_1	0	0	1	1					
2		A_2		0	0		1	0			
3		A_3	1	0	0	0		0	0	1	
4		A_n									

To make sure of the quality of information system, we define some logic-functional rules reflecting the coherence, the reliability and the comprehensiveness of the analysis by an information matrix in which the rows are relating to activities and the columns to information.

In order to become the exploitation of the information matrix more comfortable, we define the Method of Representation of the Information by Objectives (MRIO) [31] inspiring of the SADT method (Structured Analysis Design Technique) and we define its tools.

5. RESULTS OF THE SYSTEM ANALYSIS AND CORRECTIVES ACTIONS

In this part, we present the results of the system analysis of a TPP whose objective is to determine the possible reasons of the degradation of the TPP heat rate. In fact, all events that are appropriate (preventive or corrective maintenance, exploitation, in conformity of modification, related to working, work stops...) are consigned on GMAO in the TPP in order to constitute the historic and to permit the traceability. This historic has been consulted in the goal to bring a more for the possible problems research.

The objectives tree (presented in linear form) presents ten specific objectives enabling to lead the global objective (GO): TPP heat rate losses identified.

These specific objectives are: Boiler losses identified; Condensate/FW system losses identified; Circulating water system losses identified; Turbine losses identified; Steam conditions losses identified; Electrical auxiliary losses identified; Steam auxiliary losses identified; Fuel handling losses identified; Heat losses identified; Cycle isolation losses identified.

Table 2 presents the OOPP analysis of the TPP heat rate losses.

The final production of the application of the OOPP method enabled us to answer clearly to the question « what? ». Then, we presented the different results of the OOPP analysis to enable us identifying the different losses at the level of the TPP and to improve the TPP heat rate.

Figures 7 presents the objectives analysis related to the losses at the level of the boiler.

For example, the result R1.1 is decomposed in three intermediary results: Quality of the fuel identified; Moisture in air identified; Tube leaks losses identified.

Figures 8 presents the objectives analysis related to the losses at the level of the steam turbine.

Figures 9 presents the objectives analysis related to the losses at the level of the Electrical auxiliary.

Table 3 presents some performance parameters that measure how well the TPP is doing its job in producing electricity. Then, we present in the last column some corrections actions that should be made to improve thermodynamic efficiency and to improve TPP's overall performance.

Table 2. OOPP analysis.

N°	Code	Activity
1	GO	TPP heat rate losses identified
2	SO1	Boiler losses identified
3	SO2	Condensate/FW system losses identified
4	SO3	Circulating water system losses identified
5	SO4	Turbine losses identified
6	SO5	Steam conditions losses identified
7	SO6	Electrical auxiliary losses identified
8	SO7	Steam auxiliary losses identified
9	SO8	Fuel handling losses identified
10	SO9	Heat losses identified
11	SO10	Cycle isolation losses identified

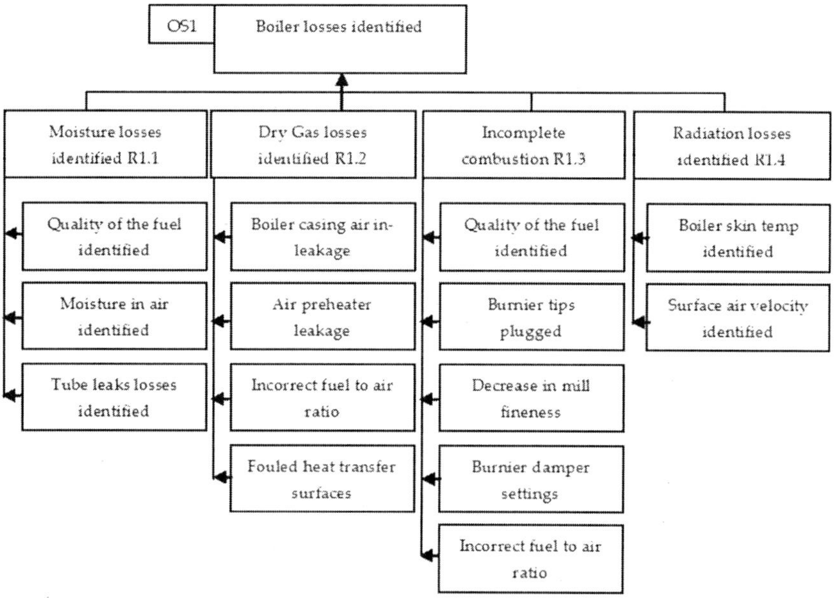

Figure 7. Objective tree of the boiler losses.

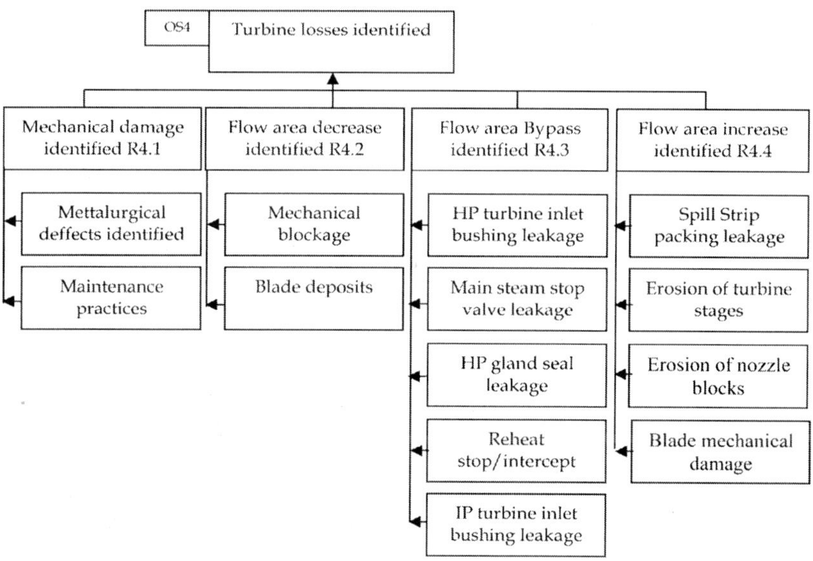

Figure 8. Objective tree of the Turbine losses.

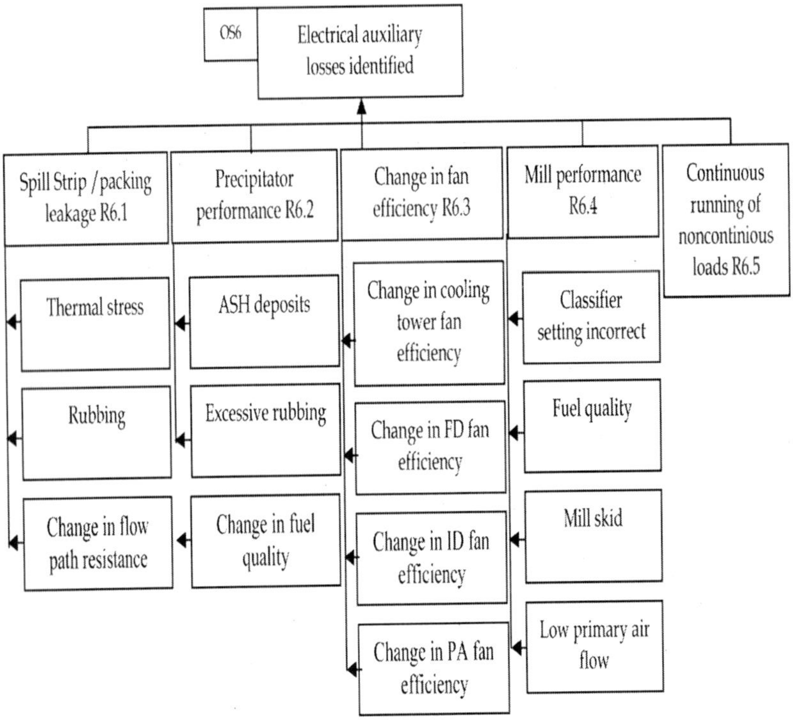

Figure 9. Objective tree of the Electrical auxiliary losses.

6. CONCLUSION

The performance of a TPP will begin to decline as the thermal power plant (TPP) begins to age. A good performance program will be able to identify these losses of the degradation of the heat rate. A more accurate knowledge of TPP heat rates can improve economic dispatching costs and ensure that profits are maintained on a daily basis.

In fact, the performance parameters measure how well the TPP is doing its job in producing electricity. Decisions should not necessarily be made only to improve thermodynamic efficiency, but rather to improve TPP's overall performance.

In this paper, we presented an exploration of the ways permitting the improvement of the TPP heat rate. This is why we presented a practical case of a TPP in Tunisia. The objective is to determine the possible causes generating losses and provoking the degradation of the TPP heat rate while using a system analysis method.

Table 3. Diagnosis of performance parameters and corrective actions.

Parameter	Cause	Test of confirmation	Corrective action
Pressure steam of the admission turbine very high	Instrument error	To verify with the redundant measures To compare with pressures in relation as (pressure exit over heater, ball)	To calibrate instruments
	No consistency with the order point	To verify if floodgates of control of the turbine are opened completely and the pressure remains even high: the coefficient of evaporation debit in the furnace is very low	To verify the order point of the pressure entrance turbine; To see the regulation system
Temperature steam of the admission turbine very low	Instrument error	To verify with the redundant measures To compare with temperatures in relation as the difference with the temperature steam admission turbine must be between 0 and 15°C	To calibrate instruments
	Pressure stem admission turbine high	Pressure steam SH is a performance parameter	
	Debit injection SH	To verify the debit of injection: the debit of injection is a performance parameter	
	Encrassement of exchange surfaces SH	Test integrity of the furnace: factor of tube encrassement	To eliminate the encrassement
	Low excess of air	To verify the excess of air: the excess of air is a performance parameter	
	Order point low	To verify the order point	To increase the value of the order point
Debit injection SH very high	Instrument error	To verify with the redundant measures ; To calculate the debit from the temperature entrance and exit heating	To calibrate instruments
	Order point of steam temperature very low	To verify the order point steam temperature	To increase the value of the order point

Parameter	Cause	Test of confirmation	Corrective action
	Debit injection on manual order point very high	To see system of regulation	To adjust the station of control of injection debit
	Very high air excess	To verify the excess of air: (performance parameter)	
	Low water temperature:		
	a- by-pass HP heating	To verify temperature water before and after by-pass HP	To close floodgate by-pass HP heating device To eliminate the flights
	B High TD heating HP	To verify TD heating HP : (performance parameter)	
	Very low pressure steam admission turbine	To verify pressure steam admission turbine	
	Flight floodgate control of injection	To close the floodgate of insulation ; To verify if the conduct is hot	To repair the floodgate
Debit injection RH very high	Instrument error	To calculate the debit of steam from the temperature of entrance and exit	
	Very low order point of temperature steam	To verify the order point	To increase the value of the order point
	Very high debit injection on manual order point	To verify the control station	To adjust the control station
	Very high air excess	To verify the air excess (performance parameter)	
	Low water temperature:		
	a- by-pass heating HP	To verify temperature water before and after by-pass HP	To close floodgate by-pass HP heating ; To eliminate the flights
	B High TD heating HP	To verify TD heating HP : (performance parameter)	

To reach this objective, System Analysis methods seem to be a promising way because the major advantage of these kinds of methods is due to the concept of hierarchy activity. These methods permit the complexity of a system to be overcome. In this paper, the application of the OOPP method on a real system, a thermal power plant in Tunisia generates a source of useful information for determining of the possible losses at the level of a TPP. So, research into the application of System Analysis methods must be intensified in order to solve several difficulties and to improve their efficiency.

REFERENCES

1. Heat Rate Improvement Reference Manual, EPRI, Palo Alto, CA, TR-109546, Jully1998.

2. Heat Rate Improvement Guidelines for Existing Fossil Plants, Electric Power Research Institute, Palo Alto, CA: May 1986. Report CS-4554.

3. A. Vitaly, Alternative trends in development of thermal power plants, Applied Thermal Engineering, 28, Issues 2-3, 2008, pp. 190-194.

4. V. Slobodan; P. Nikola and D. Željko; Power Electronics Solution to Dust Emissions from Thermal Power Plants, Serbian journal of electrical engineering, Vol.7. N°2, 2010.

5. M.N. Lakhoua, SCADA applications in thermal power plants, International Journal of the Physical Sciences, vol.5, N°7, 2010, pp. 1175-1182.

6. M. T. Khadir and S. Klai, A steam turbine diagnostic maintenance system based on an evolutive domain ontology, International Conference on Machine and Web Intelligence (ICMWI), 2010, pp. 360-367.

7. M. Yufeng and L. Yibing, An Hongwen; Statistical analysis of steam turbine faults, International Conference on Mechatronics and Automation (ICMA), 2011, pp. 2413- 2417.

8. H. Qing, D. Dongmei and L. Hong , Research on Web System of Intelligent Diagnosis for Steam Turbine, Chinese Control Conference (CCC2006), 2006, pp. 1271- 1275.

9. J. Klure-Jensen and R. Hanisch, Integration of steam turbine controls into power plant systems, IEEE Transactions on Energy Conversion, Vol.6, Issue: 1, 1991, pp. 177- 185.

10. L.N. Bize and J.D. Hurley, Frequency control considerations for modern steam and combustion turbines, IEEE Power Engineering Society 1999 Winter Meeting, Vol.1, 1999, pp. 548- 553.

11. I.C. Report, Bibliography of Literature on Steam Turbine-Generator Control Systems, IEEE Transactions on Power Apparatus and Systems, Volume: PAS-102, Issue: 9, 1983, pp. 2959- 2970.

12. W. Jianmei, C. Kai and M. Xinqiang, Optimization Management of Overhaul and Maintenance Process for Steam Turbine, ICIII '08, Vol .3, 2008, pp. 244- 247.

13. Spécification du calcul de performance ANSALDO Energia 4150 AO VVH I 082.

14. Norm ASME PTC 4.1- Steam generating units.

15. FDX 60-000, Normalisation Française, Mai 2002.

16. R.M. Roth, W.C.I. Wood and A.Delphi, Approach to acquiring knowledge from single and multiple experts, Conference on Trends and Directions in Expert Systems, 1990.

17. M. Parent, R.B. Gallupe, W.D. Salisbury and J.M. Handelman, Knowledge creation in focus groups: can group technologies help? Information & Management 38 (1), pp. 47-58, 2000.

18. H.K. Jain, M. R. Tanniru and B. Fazlollahi, MCDM approach for generating and evaluating alternatives in requirement analysis. Information Systems Research 2 (3), pp. 223-239, 1991.

19. C. Stylianou, R. L. Kumar and M. J. Khouja, A Total Quality Management-based systems development process. The DATA BASE for Advances in Information Systems 28 (3), pp. 59-71, 1997.

20. K. Schoman, D.T. Ross, Structured analysis for requirements definition, IEEE Transaction on Software Engineering 3 (1), pp. 6-15, 1977.

21. The Logical Framework Approach (LFA): Handbook for objectives-oriented planning, Norad, Fourth edition, 1999.

22. ZOPP: An Introduction to the Method, COMIT Berlin, May 1998.

23. GTZ, Methods and Instruments for Project Planning and Implementation, Eschborn, Germany, 1991.

24. AGCD, Manuel pour l'application de la «Planification des Interventions Par Objectifs (PIPO)», 2ème Edition, Bruxelles, 1991.

25. P. Gu and Y. Zhang, OOPPS: an object-oriented process planning system, Computers & Industrial Engineering, Vol.26, Issue 4, pp. 709-731, 1994.

26. S. H. Peter., Planning and change: a Cambodian public health case study, Social Science & Medicine, 51, pp.1711-1722, 2000.

27. K. Peffers, Planning for IS applications: a practical, information theoretical method and case study in mobile financial services, Information & Management, Vol.42, Issue 3, pp. 483-501, 2005.

28. S. Killich and H. Luczak, Support of Interorganizational Cooperation via TeamUp at Internet-Based Tool for Work Groups, Proceedings of the 6th internationally Scientific Conference, Berchtesgaden, May 22-25, Berlin, 2002.

29. M. Annabi, PIPO étendue : Méthode Intégrée de Spécification, de Développement et d'Implémentation de Projet (MISDIP), International conference on Sciences and Techniques of Automatic control and computer engineering STA'2003, Sousse, 2003.

30. M.N. Lakhoua, Refining the objectives oriented project planning (OOPP) into method of informational analysis by objectives, International Journal of the Physical Sciences, Vol.6(33), pp. 7550 - 7556, 2011.

31. Lakhoua, M. N. & Ben Jouida T., Refining the OOPP into Method of Representation of the Information by Objectives, International Transactions on Systems Science and Applications, Vol. 7, No. 3/4, December 2011, pp. 295-303.

Chapter 2

An Approach to Analyse Energy and Exergy Analysis of Thermal Power Plants: A Review

Vundela Siva Reddy[1], Subhash Chandra Kaushik[1], Sudhir Kumar Tyagi[2], Naraya Lal Panwar[1]

[1]Centre for Energy Studies, Indian Institute of Technology Delhi, New Delhi, India; [2]School of Infrastructure Technology & Resource Management, Shri Mata Vaishno Devi University, Katra (J&K), India.

ABSTRACT

In this paper, a thermodynamic analysis of a coal based thermal power plant and gas based cogeneration power plant has been carried out. The energy and exergy analysis has been studies for the different components of both power plants. The paper analyses the information available in the open literature regarding energy and exergy analysis on high temperature power plant has been included. A comprehensive literature review on thermal power plants, especially boiler in coal base thermal power plants and combustion chamber in gas-steam cogeneration has been included. Finally, explaining the procedure of analysis of thermal power plant systems by exegetical approach.

KEYWORDS

Exergy, Energy, Rankine Cycle

1. INTRODUCTION

The name thermodynamics comes from the Greek words therme (heat) and dynamis (power), which is most expressive of the conversion from heat into power [1]. Now adays same name is broadly interpreted to include all aspects of energy and energy transformations, including power generation, refrigeration, and relationships among the properties of matter. The First Law deals with the amounts of energy of various forms transferred between the system and its surroundings and with the changes in the energy stored in the system. It treats work and heat interactions as equivalent forms of energy in transit and offers no indication about the possibility of a spontaneous process proceeding in a certain direction. The first law places no restriction on the direction of a process, but satisfying the first law does not ensure that the process can actually occur. This inadequacy of the first law to identify whether a process can take place is remedied by introducing another general principle, the second law of thermodynamics.

The exergy method of analysis is based on the Second law of thermodynamics and the concept of irreversible production of entropy. The fundamentals of the exergy method were laid down by Carnot in 1824 and Clausius in 1865. The energy-related engineering systems are designed and their performance is evaluated primarily by using the energy balance deduced from the First law of thermodynamics. Engineers and scientists have been traditionally applying the First law of thermodynamics to calculate the enthalpy balances for more than a century to quantify the loss of efficiency in a process due to the loss of energy. The exergy concept has gained considerable interest in the thermodynamic analysis of thermal processes and plant systems since it has been seen that the First law analysis has been insufficient from an energy performance stand point.

However it can specify where the process can be improved and therefore, it will signify what areas should be given consideration. The simple energy balance will not sometimes suffice to find out the system defect. In such circumstances the exergy analysis is well thought-out to be significant to locate the systems imperfections.

Recently, we had new technologies for high temperature air combustion and ultra-high temperature combined cycle. In this case, it is necessary to study the exergy analysis on combustion and thermodynamic processes, because ordinary energy analysis does not have any evaluation supported at its temperature level. If we introduce the exergy analysis against energy analysis, which is supported by this temperature level, it is clear that the high temperature energy has a greater evaluation compared with low temperature one. In this particular field of engineering, it is difficult to use the ambient temperature energy of air and water, which are widely available. When we discuss power generation, high temperature energy of 1500°C and above in combined cycle has higher conversion efficiency than that of 500-600°C in steam cycle. In a thermodynamic cycle, it is necessary to consider the combustion, heat transfer and energy conversion processes, which include many kinds of effective and invalid items. So, when considering the above mentioned processes, the exergy analysis must be introduced to analyze power generation and heat pump cycles as against energy analysis. Recently a large number of studies based on exergy analysis have been carried out by many researchers all over the world in various system applications.

2. CYCLE ANALYSIS OF COAL FIRED THERMAL POWER PLANTS

Coal based thermal power plant working base on Rankine power cycle. The ideal cycle for vapor power cycles many of the impracticalities associated with the Carnot cycle can be eliminated by superheating the steam in the boiler and condensing it completely in the condenser, as shown schematically on a T-s diagram in **Figure 1**. The cycle that results is the Rankine cycle, which is the ideal cycle for vapor power plants. The ideal Rankine cycle does not involve any internal irreversibility and consists of the following four processes:

1-2 Isentropic compression in a pump;

2-3 Constant pressure heat addition in a boiler;

3-4 Isentropic expansion in a turbine;

4-1 Constant pressure heat rejection in a condenser.

2.1. Approach to Increase the Efficiency of the Rankine Cycle

The efficiency of Rankine cycle can be improved with considering following points:

2.1.1. Lowering the Condenser Pressure

Inside the condenser steam exists as a saturated mixture at the saturation temperature corresponding to the pressure inside the condenser. Therefore, lowering the operating pressure of the condenser automatically lowers the temperature of the steam. It increases the turbine output. It cannot be lower than the temperature of the cooling medium. Because, it increases the moisture content of the steam at the final stages of the turbine and it erodes the turbine blades. Finely decreases the turbine efficiency. Luckily, this problem can be solved by reheating of steam.

2.1.2. Superheating the Steam to High Temperatures

Superheat the steam to very high temperatures before it enters the turbine. This would be the desirable solution since the average temperature at which heat is added would also increase, thus increasing the cycle efficiency. However, it has the limitation to rise the steam temperature up to metallurgic-ally safe levels.

2.1.3. Increasing the Boiler Pressure

Increasing the boiler pressure increases the thermal efficiency of the Rankine cycle, but it also increases the moisture content of the steam to unacceptable levels.

2.1.4. Reheat Rankine Cycle

Increasing the boiler pressure increases the thermal efficiency of the Rankine cycle, but it also increases the moisture content of the steam to undesirable levels. Expand the steam in the turbine in two stages, and reheat it in between. It can decrease the excessive moisture problem in turbines. The reheat cycle was introduced in the mid-1920s, but it was deserted in the 1930s. It was reintroduces single reheat in the late 1940s and double reheats in the early 1950s because of the operational complexities. The purpose of the reheat cycle is to reduce the moisture content of the steam at the final stages of the expansion process.

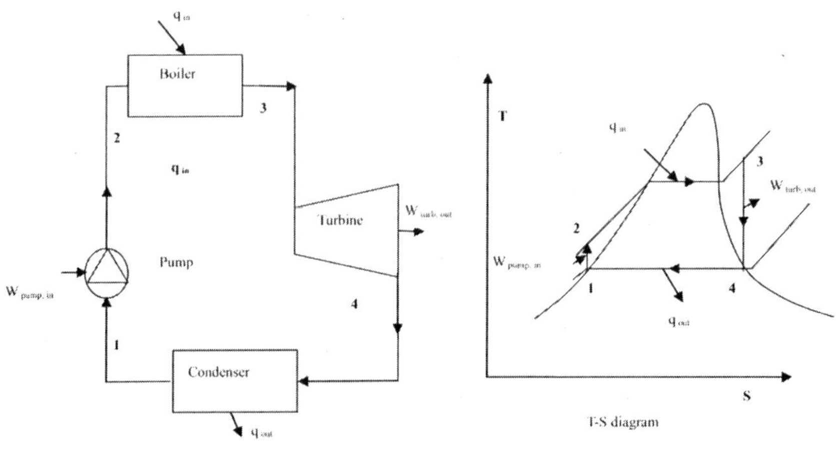

Figure 1. Schematic of rankine cycle.

2.1.5. Regenerative Rankine Cycle

To transfer heat to the feed-water from the expanding steam in a counter-flow heat exchanger built into the turbine, that is, to use regeneration. A practical regeneration process in steam power plants is accomplished by extracting/bleeding the steam from the turbine at various points. Regeneration not only improves cycle efficiency. It also helps control the large volume flow rate of the steam at the final stages of the turbine (due to the large specific volumes at low pressures).

3. CYCLE ANALYSIS OF GAS FIRED THERMAL POWER PLANTS

In this power plants toping cycle base on the Brayton cycle, operated as gas based power cycle and bottoming cycle base on the Rankine cycle. Regarding ranking cycle process we discussed in the above section. [2] The ideal cycle for gas-turbine engines: The Brayton cycle was first proposed in 1870 by George Brayton for use in the reciprocating oil-burning engine. Now a day, it is used for gas turbines only where both the compression and expansion processes take place in rotating machinery. Gas turbines usually operate on an open cycle. Fresh air at ambient conditions is drawn into the compressor; its pressure and temperature are raised. The high pressure air proceeds into the combustion chamber, where the fuel is burned at constant pressure. The resulting high-temperature gases then enter the turbine; they expand to the atmospheric pressure while producing power.

Brayton cycle, which is made up of four internally reversible processes:

1-2 Isentropic compression;

2-3 Constant-pressure heat addition;

3-4 Isentropic expansion;

4-1 Constant-pressure heat rejection.

The total reversible processes cycle shown schematically on a T-s diagram in **Figure 2**.

4. FIRST AND SECOND (EXERGY) LAW ANALYSIS

Second law is a generic term for a group of concepts that define the maximum work potential of a system, a stream of matter or a heat interaction; the state of the (conceptual) environment being used as the datum state. In an open flow system there are three types of energy transfer across the control surface namely working transfer, heat transfer, and energy associated with mass transfer or flow. [3] The work transfer is equivalent to maximum work, which can be obtained from that form of energy. The availability (Ex^Q) of heat transfer Q from the control surface at temperature T is determined from maximum rate of conversion of thermal energy to work Wmax. The Wmax is given by

$$W_{max} = Ex^{Q} = Q(1 - Ta/T) \tag{1}$$

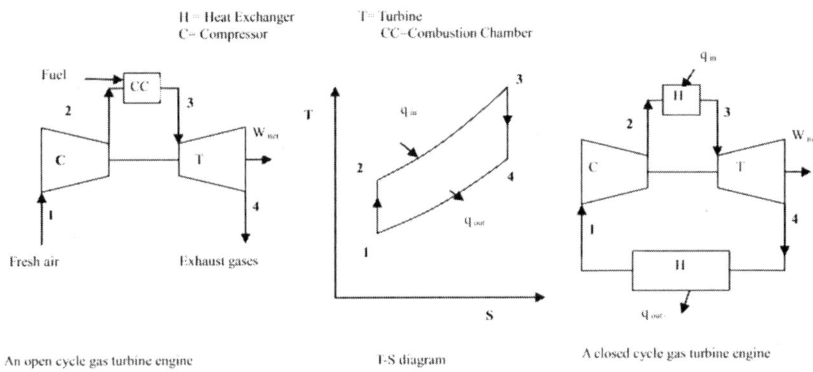

Figure 2. chematic of Brayton cycle.

Exergy of steady flow stream of matter is sum of kinetic, potential and physical exergy Ex^{ph}. The kinetic and potential energy are again equivalent to exergy. The physical specific exergy Ψ depends on initial state of matter and environmental state.

Energy analysis is based on the first law of thermodynamics, which is related to the conservation of energy. Second law analysis is a method that uses the conservation of mass and conservation of energy principles together with the entropy for the analysis, design and improvement of energy systems. Second law analysis is a useful method; to complement not to replace energy analysis. The first law of thermodynamics or energy balance for steady flow process of an open system is given by

$$\sum Q_k' - \dot{W} + \dot{m}(E_1 - E_2) = 0 \tag{2}$$

where E_1 and E_2 are respectively the energy associated with mass entering and leaving the system, Q_k is heat transfer to system from source at T_k, and

W is net work developed by the system. The Second law balance for steady flow process of an open system is given by

$$\sum Ex_{Q_k} - \dot{Ex}_W + \dot{m}(\Psi_1 - \Psi_2) - IR = 0 \tag{3}$$

where Ψ_1 and Ψ_2 are available energy associated with mass inflow and outflow are respectively, Ex_{Q_k} is available energy associated with heat transfer, Ex_W is available energy associated with work transfer and IR is irreversibility of process. The irreversibility may be due to heat transfer through finite temperature difference, mixing of fluids and mechanical friction. Second law analysis is an effective means, to pinpoint losses due to irreversibility in a real situation. The energy or first law efficiency η_I of a system or system component is defined as the ratio of energy output to the energy input of system or system component, i.e.

$$\eta_I = \frac{\text{Desired output energy}}{\text{Input energy supplied}} \tag{4}$$

The second law efficiency is defined as

$$\eta_{II} = \frac{\text{Desired output}}{\text{Maximum possible output}} \tag{5}$$

5. STUDIES CONDUCTED ON ENERGY AND EXERGY ANALYSIS OF THERMAL POWER PLANTS

Bejan [4] draw outlines the fundamentals of the methods of exergy analysis and entropy generation minimization (or thermodynamic optimization-the minimization of exergy destruction). The paper begins with a review of the concept of irreversibility, entropy generation, or exergy destruction. Examples illustrate the accounting for exergy flows and accumulation in closed systems, open systems, heat transfer processes, and power and refrigeration plants. George and Park [5] discusses how to estimate the avoidable and unavoidable exergy destruction and investment costs associated with compressors, turbines, heat exchangers and combustion chambers. This general procedure, although based on many subjective decisions, facilitates and improves applications of exergoeconomics.

5.1. Coal Fired Thermal Power Plants

Kotas [6] explained in this work the concept of exergy used to define criteria of performance of thermal plant. Ganapathy et al. [7] studied with an exergy analysis performed on an operating 50 MWe unit of lignite fired steam power plant at Thermal Power Station-I, Neyveli Lignite Corporation Limited, Neyveli, Tamil Nadu, India. The distribution of the exergy losses in several plant components during the real time plant running conditions has been assessed to locate the process irreversibility. The comparison between the energy losses and the exergy losses of the individual components of the plant shows that the maximum energy losses of 39% occur in the condenser, whereas the maximum exergy losses of 42.73% occur in the combustor. Kamate and Gangavati [8] studied exergy analysis of a heat-matched bagasse -based cogeneration plant of a typical 2500 tcd sugar factory, using backpressure and extraction condensing steam turbine is presented. In the analysis, exergy methods in addition to the more conventional energy analyses are employed to evaluate overall and component efficiencies and to identify and assess the thermodynamic losses. Boiler is the least efficient component and turbine is the most efficient component of the plant. The results show that, at optimal steam inlet conditions of 61 bar and 475°C, the backpressure steam turbine cogeneration plant perform with energy and exergy efficiency of 0.863 and 0.307 and condensing steam turbine plant perform with energy and exergy efficiency of 0.682 and 0.260.

Datta et al. [9] was presented work on exergy analysis of a coal-based thermal power plant is done using the design data from a 210 MW thermal power plant under operation in India. The exergy efficiency is calculated using the operating data from the plant at different conditions, viz. at different loads, different condenser pressures, with and without regenerative heaters and with different settings of the turbine governing. The load variation is studied with the data at 100, 75, 60 and 40% of full load. Effects of two different condenser pressures, i.e. 76 and 89 mmHg (abs.), are studied. It is observed that the major source of irreversibility in the power cycle is the boiler, which contributes to exergy destruction of the order of 60%. Part load operation increases the irreversibilities in the cycle and the effect is more pronounced with the reduction of the load. Increase in the condenser back pressure decreases the exergy efficiency. Successive withdrawal of the high pressure heaters shows a gradual increment in the exergy efficiency for the control volume excluding the boiler. Aljundi [10] was presented in this study, the energy and exergy analysis of Al-Hussein power plant in Jordan is presented. The primary objectives of this paper are to analyze the system components separately and to identify and quantify the sites having largest energy and exergy losses. In addition, the effect of varying the reference environment state on this analysis will also be presented. Energy losses mainly occurred in the condenser where 134 MW is lost to the environment while only 13 MW was lost from the boiler system. The percentage ratio of the exergy destruction to the total exergy destruction was found to be maximum in the boiler system (77%) followed by the turbine (13%), and then the forced draft fan condenser (9%). In addition, the calculated thermal efficiency based on the lower heating value of fuel was 26% while the exergy efficiency of the power cycle was 25%. For a moderate change in the reference environment state temperature, no drastic change was noticed in the performance of major components. Dai et al. [11] was done exergy analysis for each cogeneration system is examined, and a parameter optimization for each cogeneration system is achieved by means of genetic algorithm to reach the maximum exergy efficiency. The cement production is an energy intensive industry with energy typically accounting for 50-60% of the production costs. In order to recover waste heat from the preheated exhaust and clinker cooler exhaust gases in cement plant, single flash steam cycle, dual-pressure steam cycle, organic Rankine cycle (ORC) and the Kalina cycle are used for cogeneration in cement plant. The optimum performances for different cogeneration systems are compared under the same condition. The results show that the exergy losses in turbine, condenser, and heat recovery vapor generator are relatively large. Rosen [12] reported results were of energyand exergy-based comparisons of coal-fired and nuclear electrical generating stations. A version of a process-simulation computer code, previously enhanced by the author for exergy analysis, is used. Overall energy and exergy efficiencies, respectively, are 37% and 36% for the coal-fired process, and 30% and 30% for the nuclear process. The losses in both plants exhibit many common characteristics. Energy losses associated with emissions (mainly with spent cooling water) account for all of the energy losses, while emission-related exergy losses account for approximately 10% of the exergy losses. The remaining exergy losses are associated with internal consumptions. Dincer and Rosen [13] present effects on the results of energy and exergy analyses of variations in dead-state properties, and involves two main tasks: 1) examination of the sensitivities of energy and exergy values to the choice of the dead-state properties and 2) analysis of the sensitivities of the results of energy and exergy analyses of complex systems to the choice of dead-state properties. A case study of a coalfired electrical generating station is considered to illustrate the actual influences. The results indicate that the sensitivities of energy and exergy values and the results of energy and exergy analyses to reasonable variations in dead-state properties are sufficiently small. Erdem et al. [14] analyze comparatively the performance of nine thermal power plants under control governmental bodies in Turkey, from energetic and exergetic viewpoint. The considered power plants are mostly conventional reheat steam power plant fed by low quality coal. Firstly, thermodynamic models of the plants are developed based on first and second law of thermodynamics. Secondly, some energetic simulation results of the developed models are compared with the design values of the power plants in order to demonstrate the reliability. Thirdly, design point performance analyses based on energetic and exergetic performance criteria such as thermal efficiency, exergy efficiency, exergy loss, exergetic performance coefficient are performed for all considered plants in order to make comprehensive evaluations Vidal et al. [15] analysis exergy method was applied in order to evaluate the new combined cycle proposed by Goswami, using Hasan-Goswami-Vijayaraghavan parameters. This new combined cycle was proposed to produce both power and cooling simultaneously with only one heat source and using ammonia-water mixture as the working fluid. At the irreversible process two cases were considered, changing the environmental temperature. However, in order to know the performance of the new cycle at different conditions of operation, the second irreversible case was analyzed varying the rectification temperatures, the isentropic efficiency of the turbine and the return temperature of the chilled water. Exergy effectiveness values of 53% and 51% were obtained for the irreversible cycles; with heat input requirements at temperatures of 125 and 150°C. Solar collectors or waste heat are suggested as heat sources to operate the cycle.

5.1.1. Combustion

Arai et al. [3] presents an exergy analysis on combustion and energy conversion processes, which is based on the above-mentioned concept of exergy and energy supported by temperature level. When we discuss high temperature air combustion in furnace, this process shows a higher performance than that of the ambient air combustion.

5.1.2. Throttling

Luo and Wang [16] propose an approach to calculate the efficiency of throttling in which the exergy (available energy) is used to evaluate the energy conversion processes. The compressibility factor is calculated by using Peng-Robinson (P-R) equation. It is found that the largest deviation between the exergies calculated by the real gas equation and ideal gas assumption is about 1%. Because the exergy is a function of the pressure and temperature, the Joule-Thomson coefficients are used to calculate the temperature changes of throttling, based on the compressibility factors of the Soave-Redlich-Kwong (S-R-K) and P-R equations, and the temperature decreases are compared with those calculated by empirical formula.

5.1.3. Waste Emissions

Dincer and Rosen [17] studied in this paper, the concepts of exergy analysis and the linkages between exergy and environmental impact are discussed, and several issues regarding the exergies of waste emissions are addressed. Exergy is a measure of the degree of disequilibrium between a substance and its environment. The relations between several measures of environmental impact potential and exergy are investigated by comparing current methods used to assess the environmental impact potential of waste emissions and the exergy associated with those emissions.

5.2. Cogeneration and Combined Cycle Thermal Power Plants

Khaliq and Kaushik [18] presented thermodynamic methodology for the performance evaluation of combustion gas turbine cogeneration system with reheat. The energetic and exergetic efficiencies have been defined. The effects of process steam pressure and pinch point temperature used in the design of heat recovery steam generator, and reheat on energetic and exergetic efficiencies have been investigated. The power to heat ratio and second-law efficiency increases significantly with increase in process steam pressure, but the first-law efficiency decreases with the same. Results also show that inclusion of reheat provides significant improvement in electrical power output, process heat production, fuel-utilization (energetic) efficiency and second-law (exergetic) efficiency.

5.2.1. Gas Turbine Based Thermal Power Plants

Khaliq and Kaushik [19] were presented theoretical second-law approach for the thermodynamic analysis of the reheat combined Brayton/Rankine power cycle. Expressions involving the variables for specific power-output, thermal efficiency, exergy destruction in components of the combined cycle, second-law efficiency of each process of the gas-turbine cycle, and second law efficiency of the steam power cycle have been derived. It is found that the exergy destruction in the combustion chamber represents over 50% of the total exergy destruction in the overall cycle. The combined cycle efficiency and its power output were maximized at an intermediate pressure-ratio, and increased sharply up to two reheat-stages and more slowly thereafter. Chen and Tyagi [20] were presented parametric study of an irreversible cycle model of a regenerative-intercooled-reheat Brayton heat engine along with a detailed. The power output and the efficiency are optimized with respect to the cycle temperatures for a typical set of operating conditions. It is found that there are optimal values of the turbine outlet temperature, inter cooling, reheat and cycle pressure ratios at which the cycle attains the maximum power output and efficiency. But the optimal values of these parameters corresponding to the maximum power output are different from those corresponding to the maximum efficiency for the same set of operating condition. Kaushik and Tyagi [21] explained in this paper a parametric study of an irreversible regenerative Brayton heat engine with isothermal heat addition has been performed with external as well as internal irreversibilities. The external irreversibility is due to the finite temperature differences between the heat engine and the external reservoirs, while the internal irreversiblities are due to other processes, viz. non-isentropic compression and expansion processes in the compressor and turbine, respectively, and the regenerative heat loss. There is a significant improvement in the thermal efficiency (above 15%) of a Brayton cycle with isothermal heat addition over the conventional one. It is seen that the effect of the isobaric side effectiveness is rather pronounced for the power output and the corresponding thermal efficiency. Zubair [22] was examined the performance of regenerative -reheat power plants in terms of irreversibility analysis. The reduction in the irreversible losses with the addition of backward, cascade-type feed water heaters and/or a reheat option are compared with a conventional energy -balance approach. The results indicate that most of the irreversible losses occur in the boiler and that these losses are significantly reduced by the incorporation of feed water heating. The incorporation of feed water heating results in a reduction of the total irreversibility rate of the cycle by 18%. The corresponding improvement in efficiency is 12%.

5.2.2. Heat Exchangers

Yilmaz et al. [23] presented in his paper, second-law based performance evaluation criteria to evaluate the performance of heat exchangers. First, the need for the systematic design of heat exchangers using a second law-based procedure is recalled and discussed. Then, a classification of second-law based performance criteria is presented: 1) criteria that use entropy as evaluation parameter, and 2) criteria that use exergy as evaluation parameter. Both classes are collectively presented and reviewed, and their respective characteristics and constraints are given.

5.2.3. Combined Cycle Based Thermal Power Plants

The Schematic diagram of gas fired combined cycle thermal power plant is shown in **Figure 3**. Chen et al. [24] present performance analysis and optimization of an open-cycle regenerator gas-turbine power plant. The analytical formulae about the relation between power output and cycle overall pressure-ratio are derived taking into account the eight pressure-drop losses in the intake, compression, regeneration, combustion, expansion and discharge processes and flow process in the piping, the heat-transfer loss to the ambient environment, the irreversible compression and expansion losses in the compressor and the turbine, and the irreversible combustion loss in the combustion chamber. The power output is optimized by adjusting the mass-flow rate and the distribution of pressure losses along the flow path. Reddy and Mohamed [25] present work, exergy analysis of a natural gas fired combined cycle power generation unit is performed to investigate the effect of gas turbine inlet temperature and pressure ratio on exergetic efficiency for the plant and exergy destruction/losses for the components. For a fixed gas turbine inlet temperature, an optimum pressure ratio exists where the exergy destruction is less [26]. The effect of steam injection on the performance of natural gas fired combined cycle power plant is investigated. For a particular combined cycle power generation configuration, the effect of steam injection on the performance of topping cycle, and carbon dioxide emissions is conducted based on first law of thermodynamics.

Sohn et al. [27] described details of exergy-based performance characteristics of a heavy-duty gas turbine, 150MW-class GE 7F model. Results have shown that a chemical reaction in the combustor of which the exergy destruction ratio is 28.3% at full-load is one of the major sources of exergy destructions in the gas turbine it was discovered that the magnitude of exergy destruction by the cooling air in turbine stages is large enough to influence the overall turbine efficiency. The exergy destruction by the cooling air is more than half of the total exergy destruction of each cooled turbine stage Results have shown that a chemical reaction in the combustor of which the exergy destruction ratio is 28.3% at full-load is one of the major sources of exergy destructions in the gas turbine. Koroneos et al. [28] studied the exergy analysis of solar energy, wind power and geothermal energy. That is, the actual use of energy from the existing available energy is discussed. In addition, renewable energy sources are compared with the non-renewable energy sources on the basis of efficiency. Khaliq [29] was proposed conceptual trigeneration system based on the conventional gas turbine cycle for the high temperature heat addition while adopting the heat recovery steam generator for process heat and vapor absorption refrigeration for the cold production. Combined first and second law approach is applied and computational analysis is performed to investigate the effects of overall pressure ratio, turbine inlet temperature, pressure drop in combustor and heat recovery steam generator, and evaporator temperature on the exergy destruction in each component, first law efficiency, electrical to thermal energy ratio, and second law efficiency of the system.

It also indicates that maximum exergy is destroyed during the combustion and steam generation process; which represents over 80% of the total exergy destruction in the overall system Ivar et al. [30] was investigated a concept for natural-gas fired power plants with CO_2 capture using exergy analysis. Natural-gas was reformed in an auto-thermal reformer, and the CO_2 was separated before the hydrogen-rich fuel was used in a conventional combined-cycle process. A corresponding conventional combined-cycle power plant with no CO_2 capture was simulated for comparison. A base case with CO_2 capture was specified with turbine-inlet temperature of 1250°C and an air-compressor outlet pressure of 15.6 bar. In this case, the net electric-power production was 48.9% of the lower heating value of the Natural-gas or 46.9% of its chemical exergy. The effect of increased turbine-inlet temperature to 1450°C was investigated. Combining both measures, the net electric power production was increased to 53.3% of the natural-gas lower heating value or 51.1% of the natural-gas chemical exergy. On the other hand, both increased turbine-inlet temperature and the auto-thermal reformer product-feed heat exchange reduced the conversion of hydrocarbons to CO_2. Nag and Gupta [31] was down exergy analysis , to reduce thermal irreversibilitys a thermodynamic cycle was conceptualized for the bottoming part of a combined cycle by Kalina, which uses NH3- H2O mixture as the working substance. The effect of one of the key parameter, mixture concentration at turbine inlet on the cycle performance has been studied. Franco and Russo [32] presented in this present particular work point out attention to the optimization of the heat recovery steam generator (HRSG), as a first step in the analysis of the whole plant, according to a hierarchical strategy. He handles this problem adopting both a thermodynamic and a thermo economic objective function instead that with the usual pinch point method. Thermodynamic optimization has the purpose to diminish energy losses, expressed on exergy basis, while the aim of the thermo economic optimization is the minimization of a cost function, sum of the cost of exergy inefficiencies and the cost of the HRSG. Bilgen [33] presented exergetic and engineering analyses as well as a simulation of gas turbine-based cogeneration plants consisting of a gas turbine, heat recovery steam generator and steam turbine. The exergy analysis is based on the first and second laws of thermodynamics. The engineering analysis is based on both the methodology of levelized cost and the payback period.

Dincer et al. [34] presented an efficiency analysis, accounting for both energy and exergy considerations, of a design for a cogeneration-based district energy system. A case study is considered for the city of Edmonton, Canada, by the utility Edmonton Power. The energyand exergy-based results differ markedly (e.g., overall energy efficiencies are shown to vary for the three configurations considered from 83% to 94%, and exergy efficiencies from 28% to 29%, respectively). Datta et al. [35] studied thermal performance and sizing of Biomass based decentralized power generation using externally fired gas turbine can be a technically feasible option plants have been analyzed at different cycle pressure ratio (r_p = 2-8), turbine inlet temperature (1050-1350 K) and the heat exchanger cold end temperature difference (200-300 K). It is found that the thermal efficiency of the externally fired gas turbine plant reaches a maximum at an optimum pressure ratio depending upon the turbine inlet temperature and heat exchanger. For a particular pressure ratio, thermal efficiency increases either with the increase in turbine inlet temperature or with the decrease in heat exchanger cold end temperature. Sue and Chuang [36] presented in his paper the engineering design and theoretical exergetic analyses of the plant for combustion. The results show the exergy analyses for a steam cycle system predict the plant efficiency more precisely. The plant efficiency for partial load operation is lower than full load operation. Increasing the pinch points will decrease the combined cycle plant efficiency of gas turbine based power generation systems.

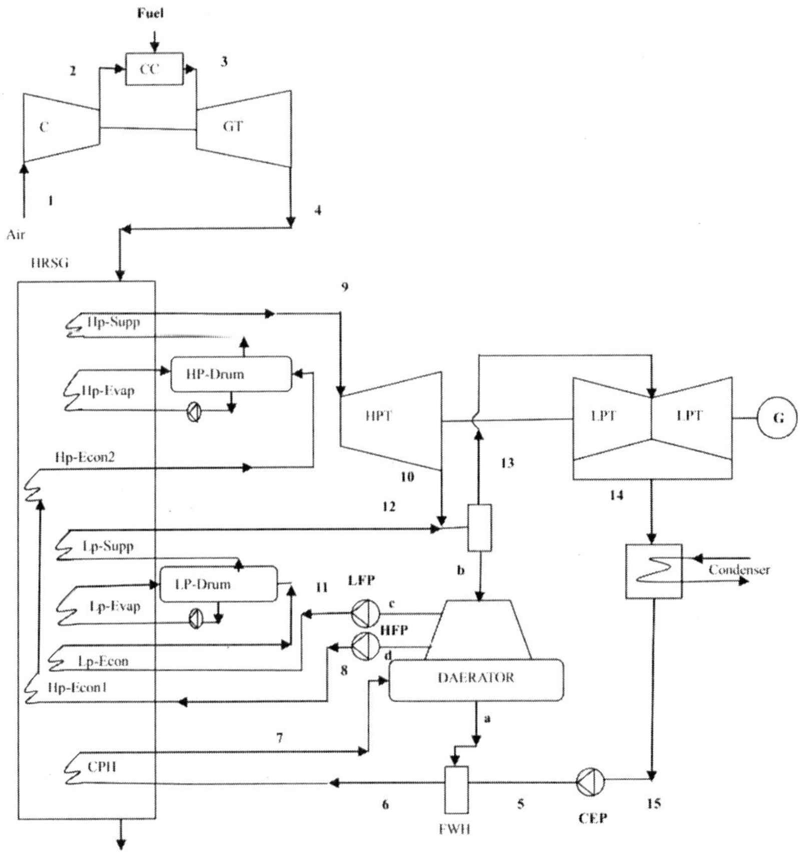

Figure 3. Schematic of gas fired combined cycle thermal power plant.

6. CONCLUSIONS

The worldwide concern about cost, environment and quick availability to meet continuous load growth will continue to enhance the adoption of gas and steam turbine engines in power systems. The escalating interest in efficient use of energy will support the adoption of cogeneration with simultaneous production of power and thermal energy. The first law analysis shows major energy loss has been found to occur in condenser. The second law (Exergy) analysis shows that combustion chamber in both steam and gas turbine thermal power plants are main source of Irreversibility. The Irreversibility in condenser is insignificant, because in the condenser the low quality energy is lost. An Exergy method of optimization gives logical solution improving the power production opportunities in thermal power plants.

REFERENCES

1. A. Yunus and A. Michael, "Thermodynamics an Engineering Approach," Tata McGraw Hill, New Delhi, 2007.

2. P. K. Nag, "Power Plant Engineering," Tata McGraw Hill, New Delhi, 2007.

3. N. Arai, H. Taniguchi, K. Mouri and T. Nakahara, "Exergy Analysis on Combustion and Energy Conversion Processes," Energy, Vol. 30, No. 2-4, 2005, pp. 111-117.

4. A. Bejan, "Fundamentals of Exergy Analysis, Entropy Generation Minimization, and the Generation of Flow Architecture," International Journal of Energy Research, Vol. 26, No. 7, 2002, pp. 545-565.

5. G. Tsatsaronis and M. Park, "On Avoidable and Unavoidable Exergy Destructions and Investment Costs in Thermal Systems," Energy Conversion & Management, Vol. 43, No. 9-12, 2002, pp. 1259-1270.

6. T. J. Kotas, "Exergy Criteria of Performance for Thermal Plant: Second of Two Papers on Exergy Techniques in Thermal Plant Analysis," International Journal of Heat and Fluid Flow, Vol. 2, No. 4, 1980, pp. 147-163.

7. T. Ganapathy, N. Alagumurthi, R. P. Gakkhar and K. Murugesan, "Exergy Analysis of Operating Lignite Fired Thermal Power Plant," Journal of Engineering Science and Technology Review, Vol. 2, No. 1, 2009, pp.123-130.

8. S. C. Kamate and P. B. Gangavati, "Exergy Analysis of Cogeneration Power Plants in Sugar Industries," Applied Thermal Engineering, Vol. 29, No. 5-6, 2009, pp. 1187- 1194.

9. A. Datta, S. Sengupta and S. Duttagupta, "Exergy Analysis of a Coal-Based 210 mw Thermal Power Plant," International Journal of Energy Research, Vol. 31, No. 1, 2007, pp. 14-28.

10. I. H. Aljundi, "Energy and Exergy Analysis of a Steam Power Plant in Jordan," Applied Thrmal Engineering, Vol. 29, No. 2-3, 2009, pp. 324-328.

11. Y. Dai, J. Wang and L. Gao, "Exergy Analyses and Parametric Optimizations for Different Cogeneration Power Plants in Cement Industry," Applied Energy, Vol. 86, No. 6, 2009, pp. 941-948.

12. M. A. Rosen, "Energyand Exergy-Based Comparison of Coal-Fired and Nuclear Steam Power Plants," Exergy, Vol. 1, No. 3, 2001, pp. 180-192.

13. I. Dincer and M. A. Rosen, "Effect of Varying Dead-State Properties on Energy and Exergy Analyses of Thermal Systems," International Journal of Thermal Sciences, Vol. 43, No. 3, 2004, pp. 121-133.

14. H. Erdem, A. V. Akkaya., A. Dagdas, S. H. Sevilgen, B. Sahin, I. Tek, C. Gungor and S. Atas, "Comparative Energetic and Exergetic Performance Analyses for CoalFired Thermal Power Plants in Turkey," International Journal of Thermal Sciences, Vol. 48, No. 11, 2009, pp. 2179-2186.

15. A. Vidal, R. Best, R. Rivero and J. Cervantes, "Analysis of a Combined Power and Refrigeration Cycle by the Exergy Method," Energy, Vol. 31, No. 15, 2006, pp. 3401- 3414.

16. Y. X. Luo and X. Y. Wang, "Exergy Analysis on Throttle Reduction Efficiency Based on Real Gas Equations," Energy, Vol. 35, No. 1, 2010, pp. 181-187.

17. I. Dincer and M. A. Rosen, "Exergy Analysis of Waste Emissions," International Journal of Energy Reserch, Vol. 23, No. 13, 1999, pp. 1153-1163.

18. A. Khaliq and S. C. Kaushik, "Thermodynamic Performance Evaluation of Combustion Gas Turbine Cogeneration System with Reheat," Applied Thermal Engineering, Vol. 24, No. 13, 2004, pp. 1785-1795.

19. A. Khaliq and S. C. Kaushik, "Second-Law Based Thermodynamic Analysis of Brayton/Rankine Combined Power Cycle with Reheat," Applied Energy, Vol. 78, No. 2, 2004, pp. 179-197.

20. G. M. Chen, S. K. Tyagi, Q. Wang and S. C. Kaushik, "A New Thermoeconomic Approach and Parametric Study of an Irreversible Regenerative Brayton Refrigeration Cycle," International Journal of Refrigeration, Vol. 29, No. 7, 2006, pp. 1167-1174.

21. S. C. Kaushik, S. K. Tyagi and M. K. Singhal, "Parametric Study of an Irreversible Regenerative Brayton Cycle with Isothermal Heat Addition," Energy Conversion & Management, Vol. 44, No. 12, 2003, pp. 2013-2025.

22. S. M. Zubair and M. A. Habib, "Second-Law-Based Thermodynamic Analysis of Regenerative-Reheat RankineCycle Power Plants," Energy, Vol. 17, No. 3, 1992, pp. 295-301.

23. M. Yilmaz, O. N. Sara and S. Karsli, "Performance Evaluation Criteria for Heat Exchangers Based on Second Law Analysis," Exergy, Vol. 1, No. 4, 2001, pp. 278-294.

24. L. G. Chen, Y. Li, F. R. Sun and C. Wu, "Power Optimization of Open-Cycle Regenerator Gas-Turbine PowerPlants," Applied Energy, Vol. 78, No. 2, 2004, pp. 199- 218.

25. B. V. Reddy and K. Mohamed, "Exergy Analysis of Natural Gas Fired Combined Cycle Power Generation Unit," International Journal of Exergy, Vol. 4, No. 2, 2007, pp. 180-196.

26. B.V. Reddy and I. E. Alaefour, "Performance Simulation of a Natural Gas Fired Combined Cycle Power Generation System," 19th National & 8th ISHMT-ASME Heat and Mass Transfer Conference, Hyderabad, 2008.

27. J. L. Sohn, T. W. Song, J. H. Kim, T. S. Kim and S. T. Ro, "Exergy-Based Performance Analysis of the Heavy-Duty Gas Turbine in Part-Load Operating Conditions," Exergy, Vol. 2, No. 2, 2002, pp. 105-112.

28. C. Koroneos, T. Spachos and N. Moussiopoulos, "Exergy Analysis of Renewable Energy Sources," Renewable Energy, Vol. 28, No. 2, 2003, pp. 295-310.

29. A. Khaliq, "Exergy Analysis of Gas Turbine Trigeneration System for Combined Production of Power Heat and Refrigeration," Refrigeration, Vol. 32, No. 3, 2009, pp. 534-545.

30. I. S. Ertesvag, H. M. Kvamsdal and O. Bolland, "Exergy Analysis of a Gas-Turbine Combined-Cycle Power Plant with Precombustion Co_2 Capture," Energy, Vol. 30, No. 1, 2009, pp. 5-39.

31. P. K. Nag and A. V. S. S. K. S. Gupta, "Exergy Analysis of the Kalina Cycle," Applied Thermal Engineering, Vol. 18, No. 6, 1998, pp. 427-439.

32. A. Franco and A. Russo, "Combined Cycle Plant Efficiency Increase Based on the Optimization of the Heat Recovery Steam Generator Operating Parameters," International Journal of Thermal Sciences, Vol. 41, No. 9, 2002, pp. 843-859.

33. E. Bilgen, "Exergetic and Engineering Analyses of Gas Turbine Based Cogeneration Systems," Energy, Vol. 25, No. 12, 2000, pp. 1215-1229.

34. I. Dincer, M. A. Rosen and N. L. Minh, "Efficiency Analysis of a Cogeneration and District Energy System," Applied Thermal Engineering, Vol. 25, No. 1, 2005, pp. 147-159.

35. A. Datta, R. Ganguly and L. Sarkar, "Energy and Exergy Analyses of an Externally Fired Gas Turbine (Efgt) Cycle Integrated with Biomass Gasifier for Distributed Power Generation," Energy, Vol. 35, No. 1, 2010, pp. 341-350.

36. D.-C. Sue and C.-C. Chuang, "Engineering Design and Exergy Analyses for Combustion Gas Turbine Based Power Generation System," Energy, Vol. 29, No. 8, 2004, pp. 1183-1205.

Chapter 3

Application of Functional Analysis Techniques and Supervision of Thermal Power Plants

M.N. Lakhoua[1]

[1] Member IEEE Laboratory of Analysis and Command of Systems (LACS)ENIT, Le Belvedere, Tunis,, Tunisia

1. INTRODUCTION

Supervision consists of commanding a process and supervising its working [1]. To achieve this goal, the supervisory system of a process must collect, supervise and record important sources of data linked to the process, to detect the possible loss of functions and alert the human operator.

The main objective of a supervisory system is to give the means to the human operator to control and to command a highly automated process [2]. So, the supervision of industrial processes includes a set of tasks aimed at controlling a process and supervising its operation.

An automatic supervisory system is a traditional supervisory system, that is to say, a system which provides a hierarchical list of alarms generated by a simple comparison with regard to thresholds [3]. The information synthesis system manages the presentation of information via any support (synoptic, console, panel, etc.) to the human operator.

Today, the supervision of production systems is more and more complex to perform, not only because of the number of variables always more numerous to monitor but also because of the numerous interrelations existing between them, very difficult to interpret when the process is highly automated[4].

The challenge of the future years is based on the design of support systems which let an active part to the supervisory operators by supplying tools and information allowing them to understand the running of production equipment. Indeed, the traditional supervisory systems present many already known problems. First, whereas sometimes the operator is saturated by an information overload, some other times the information under load does not permit them to update their mental model of the supervised process [5].

Moreover, the supervisory operator has a tendency to wait for the alarm to act, instead of trying to foresee or anticipate abnormal states of the system. So, to avoid these perverse effects and to make operator's work more active, the design of future supervisory systems has to be human centred in order to optimize Man-Machine interactions [6].

It seems in fact important to supply the means to this operator to perform his own evaluation of the process state. To reach this objective, Functional Analysis seems to be a promising research method. In fact, allowing the running of the production equipment to be understood, these techniques permit designers to determine the good information to display through the supervisory interfaces dedicated to each kind of supervisory task (monitoring, diagnosis, action, etc.). In addition, Functional Analysis techniques could be a good help to design support systems such as alarm filtering systems [7].

By means of a significant example, the objective of this paper is to show interests of the use of Functional Analysis (FA) techniques such as SADT (Structured Analysis Design Techniques) and SA-RT (Structured Analysis Real Time) for the design of supervisory systems. An example of a SCADA system of a thermal power plant (TPP) is presented. The next section briefly describes the characteristics of a SCADA system and the problems linked to its design. Next, the interests of using FA in the design steps are developed. In Section 3, after presenting concepts of SADT and SA-RT, these methods are applied to a SCADA system of a TPP. The last section presents a discussion about the advantages and inconveniences of FA techniques and some examples of SCADA applications in a TPP.

2. PRESENTATION OF SUPERVISORY CONTROL AND DATA ACQUISITION SYSTEM

The SCADA term (supervisory control and data acquisition) refers to a system that collects data coming from different sensors of an industrial or other process [8] [9] [10], these sensors can beings installed in the same site or distant (several Km), the introverted data are treated by an unit called processor power station (CPU, PCU, PC...), results are sent in real time to the Men / Machine interfacing that can be a computer with its peripherals (See Figure 1).

The SCADA system of the TPP of Radès (in Tunisia) orders and classifies all data for [11] [12]:

- Instantaneous impression.
- Visualization on screen using data tables and tabular diagrams.
- Registration of instantaneous exchanges of numeric and analogical data.
- Instantaneous calculation for example corrections of gas debits, direct middle specific consumption, middle values.
- Storage of the analogical information of the process.
- Calculation of outputs and losses of the process.
- Surveillance of the SOE signals (entrances rapid contact 1ms)
- Interfacing interactive Men / Machine for the surveillance of the system and the conduct of processes (tabular, curves view of alarm).

The objective of the SCADA system of the TPP of Radès is to collect data instantaneously (ON LINE) of their sites and to transform them in numeric data. This centralized supervision allows operators, since the room of control of the TPP, to control facilities in their domain of exploitation and the different types of incidents.

The SCADA system of the TPP of Radès is equipped of three networks of communication:

- Field bus, 5 Mbits, permitting to do exchanges of the numeric data of the entrance card / exits (FBM) toward the central system (CP) via modules of communication (FCM);
- Node bus, 10 Mbits, permitting to do exchanges of the numeric data of the central system (CP) via modules of communication (DNBT) toward the Men/Machine interfacing (workstations);

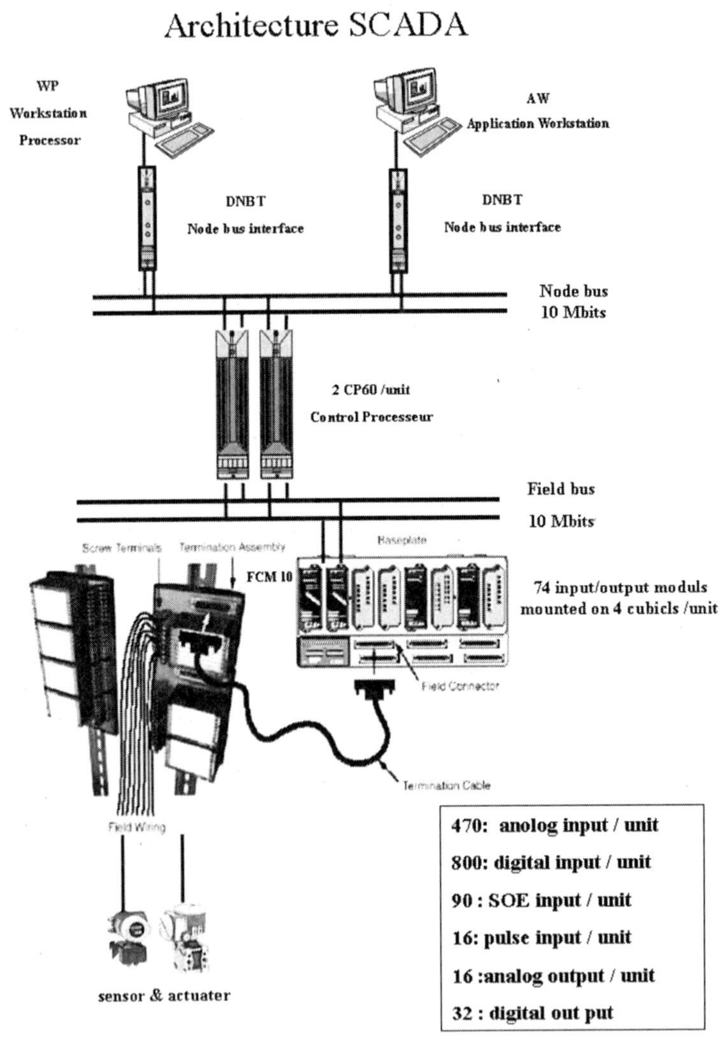

Figure 1. SCADA system of the TPP.

Figure 2 presents the different links between the CP60, FCM and FBM blocks.

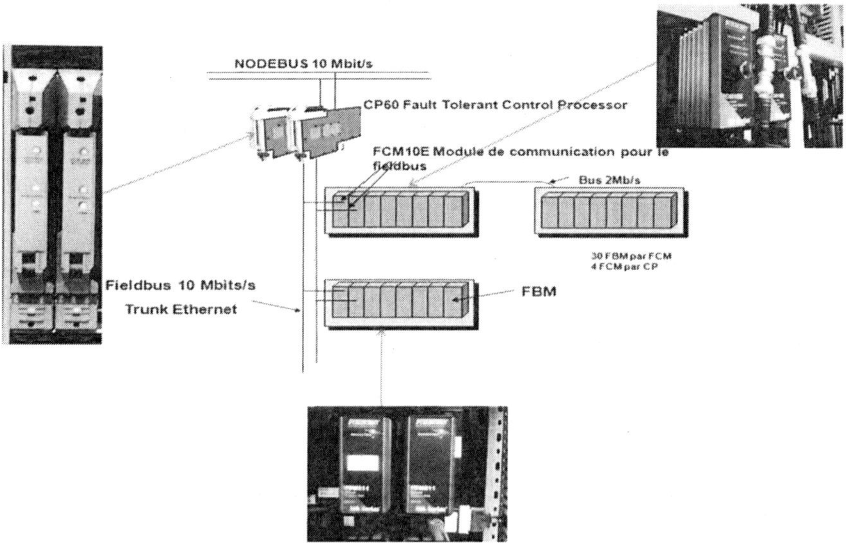

Figure 2. The CP60/FCM/ FBM links.

3. FUNCTIONAL ANALYSIS OF A SCADA SYSTEM

There are many methods that have been used to enhance participation in Information System planning and requirements analysis [13]. We present the application of SADT and SA-RT methods on a SCADA system.

The SADT method represents attempts to apply the concept of focus groups specifically to information systems planning, eliciting data from groups of stakeholders or organizational teams. SADT is characterized by the use of predetermined roles for group/team members and the use of graphically structured diagrams. It enables capturing of proposed system's functions and data flows among the functions [14].

SADT, which was designed by Ross in the 1970s [15], was originally destined for software engineering but rapidly other areas of application were found, such as aeronautic, production management, etc.

For SADT diagrams or function boxes, we will consider two events to be representing the activation states of the activities. The first event represents the instant when the activity is triggered off, and the second event represents the ending instant.

Figure 3 presents the Actigram and Datagram of the SADT model.

The boxes called ICOM's input-control-output-mechanisms are hierarchically decomposed. At the top of the hierarchy, the overall purpose of the system is shown, which is then decomposed into components-subactivities. The decomposition process continues until there is sufficient detail to serve the purpose of the model builder.

Figures 4 and 5 present respectively the A-0 and the A0 levels of the SADT model of a SCADA system [16] [17].

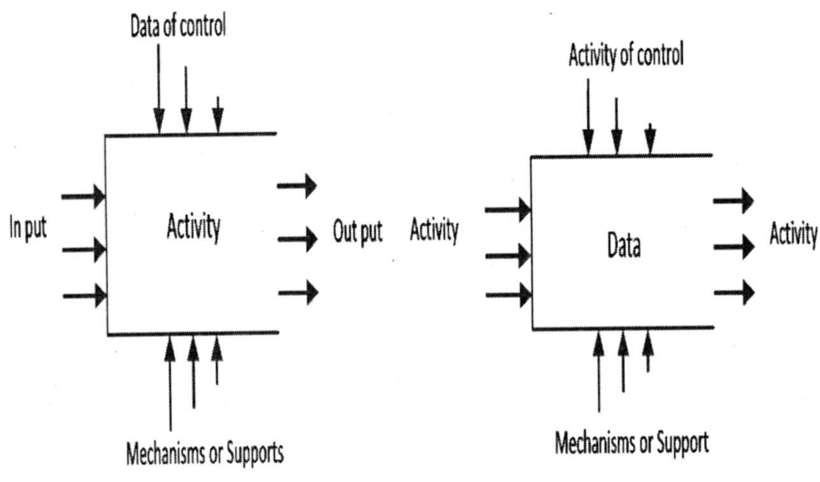

Figure 3. Organization of the SADT model [19].

SADT is a standard tool used in designing computer integrated manufacturing systems, including flexible manufacturing systems. Although SADT does not need any specific supporting tools, several computer programs implementing SADT methodology have been developed. One of them is Design: IDEF, which implements IDEF0 method. SADT: IDEF0 represents activity oriented modeling approach.

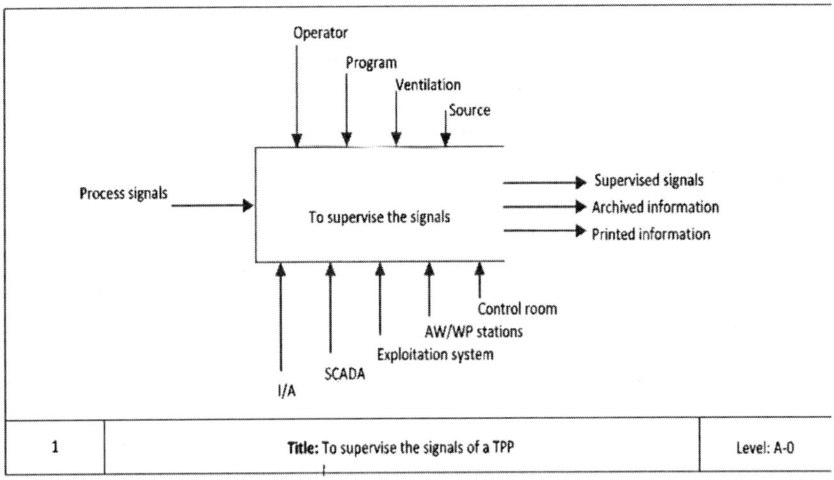

Figure 4. A-0 level of the SADT model [16].

Among the graphical methods most commonly used in industry, two of the leading methods are SA-RT and Statecharts. SA-RT is a short name for Structured Analysis Methods with extensions for Real Time [18]. The model is represented as a hierarchical set of diagrams that includes data and control transformations (processes). Control transformations are specified using State Transition diagrams, and events are represented using Control Flows. The other graphical and state based paradigm for specification of real time systems is Statecharts. The system is represented as a set of hierarchical states instead of processes. Each state can be decomposed into sub states and so on. The statecharts notation is more compact than the SA-RT notation and has been formally defined.

Structured Analysis for Real-Time Systems, or SA-RT, is a graphical design notation focusing on analyzing the functional behavior of and information flow through a system [19]. SA-RT, which in turn is a refinement of the structural analysis methods originally introduced by Douglass Ross and popularized by Tom DeMarco in the seventies, was first introduced by Ward and Mellor in 1985 and has thereafter been refined and modified by other researchers, one well-known example being the Hatley and Pirbhai proposal (See Figure 6).

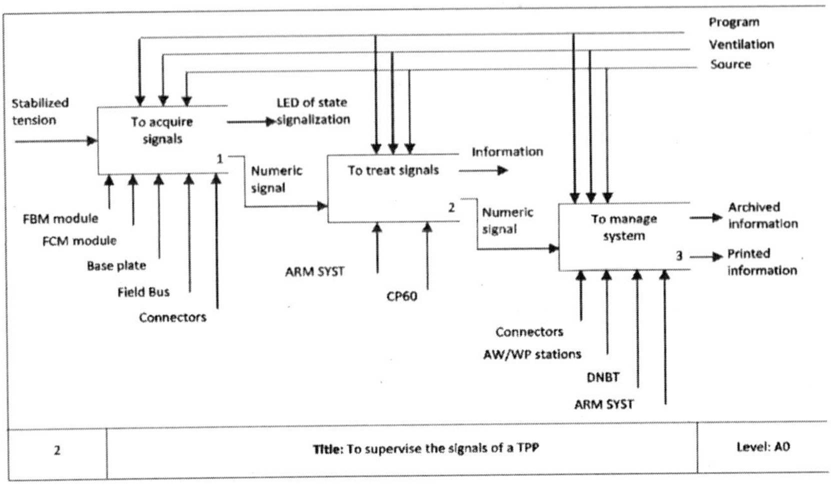

Figure 5. A0 level of the SADT model [16].

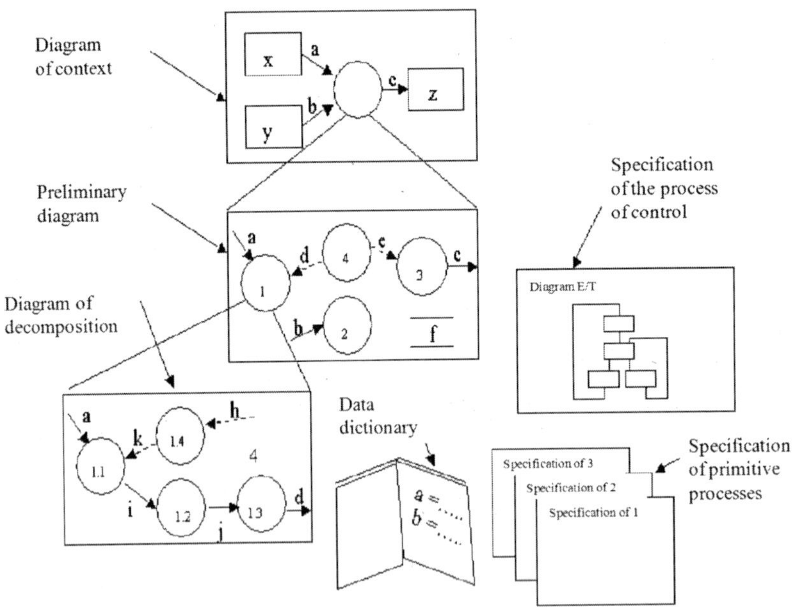

Figure 6. Organization of an SA-RT model [19].

Thus, SA-RT is a complex method for system analysis and design. This is one of the most frequently used design method in technical and real-time oriented applications adopted by various Case-Tools. It is a graphical, hierarchical and implementation independent method for top-down development.

SA-RT method enables us to identify an entrance and an exit of data in an algorithm or a computer program. It is divided in three modules: Diagram of Context, Data Flows Diagram and Control Flows Diagram. Every module includes in its graphic interpretation different symbols.

So, the SA-RT model is composed exclusively of diagrams [20]. It starts with the main process 'To supervise the signals of the TPP' (Figure 7). Then, this process is broken into a preliminary data flows diagram composed of three processes. We continue the decomposition of the processes until the last decomposition level has been reached (levels DFD1, DFD2 and DFD3).

For this example, we present the control flow diagram of the application of the SCADA system in a hydrogen station (Figure 8).

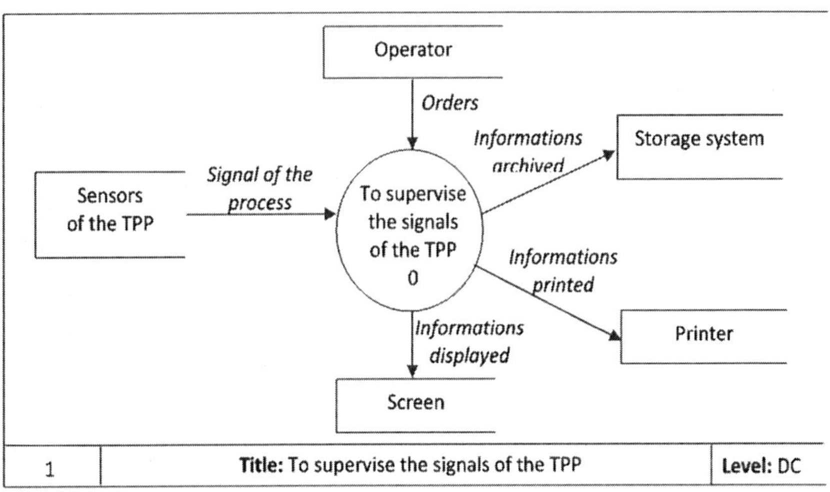

Figure 7. Diagram of Context of the SCADA system [20].

In order to give a detailed vision of the control of the hydrogen station, we present on Figure 9 an example of a State/Transition diagram.

Compared to the results given by the SADT method, the SA-RT method allows a functional as well as a temporal analysis.

The possible uses for the SADT and the SA-RT models are the design of a monitoring display and a diagnosis display. For the design of a monitoring display, the A0 level of the SADT model or the preliminary data flow diagram of the SA-RT model supplies a global view of the system. Indeed, information relative to each function represented through this level should appear in a monitoring display.

For the design of hierarchical diagnosis displays, each actigram of the SADT model or data flows diagram of the SA-RT model constitutes a vision at a given abstraction level. So, each of these actigrams gives a less or more detailed vision. In function of the objectives defined by the designer for each display, a particular actigram or data flows diagram can supply the required information.

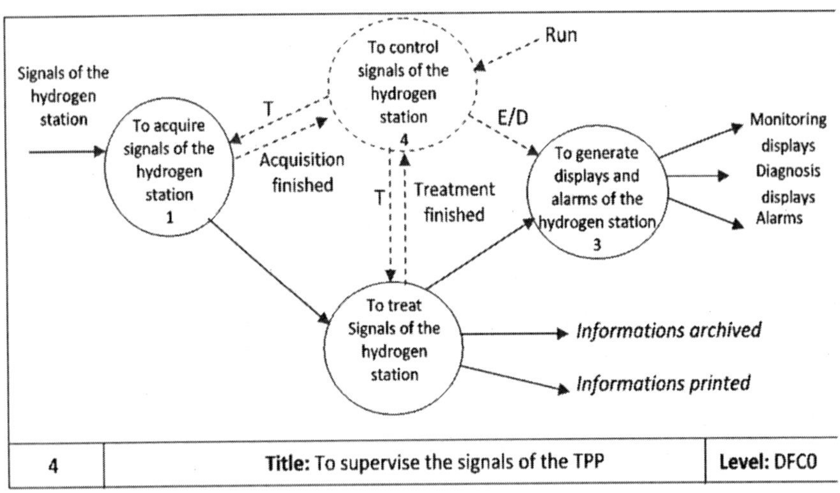

Figure 8. Control Flows Diagram of the SCADA system [20].

Finally, this application of the FA techniques on the SCADA system of a TPP shows briefly the interests of the SADT and SA-RT methods in the design of supervisory systems.

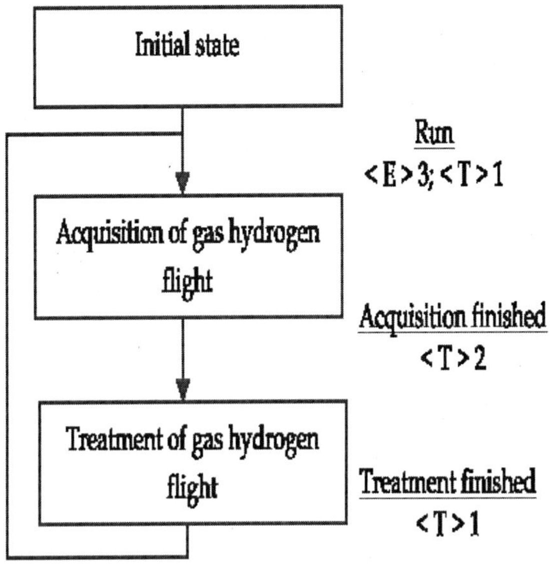

Figure 9. State/Transition diagram of the SCADA system [20].

4. APPLICATION OF THE SUPERVISION OF A HYDROGEN STATION OF A TPP

The objective of this application is to show interests of the use of a SCADA system in a hydrogen station of a TPP. In fact, the cooling by hydrogen has been adopted for turbo-alternators in 1926. This technique has been used for the interior cooling of drivers while doing circulating of the fluid in their conducts, putting the fluid in contact with materials in which the heat is produced.

Hydrogen is an odorless, colorless, very light gas (more that air) and composed of two atoms of hydrogen. It possesses a high gravimetric energizing power: 120 MJ/kgs compared to oil (45 MJ/kgs), to the methanol (20 MJ/kgs) and to the natural gas (50 MJ/kgs). However it is as the lightest gas (2,016g/mol H_2), of where a weak volumetric power: 10,8MJ/m^3 facing the methanol (16 MJ/m^3), natural gas (39.77 MJ/m^3). It puts a real problem of storage and transport: that it is for the utilization of hydrogen in a vehicle or for the transport in pipeline, in truck, it is the volumetric density that imports. The volumetric energizing density of H2 is not interesting that to the state liquid tablet either (700 bars).

The fact that a mixture of hydrogen and air is an exploding mixture on a large range of proportions, the machine and the procedure of utilization specified is conceived so that no exploding mixture can occur in the normal conditions of working.

The aforesaid unforeseen conditions cannot present themselves that during the replacement. However the pressure of gas being nearly equal to the air pressure in this condition, the intensity of explosion can take place doesn't pass to the more of 7bar.

The purity of hydrogen H_2 in an alternator is always maintained superior to 95% until 98% and when it decreases to 90% an alarm is given out to the panel of the local cupboard as well as the room of control. Preventing gas H_2 intern to form an exploding the mixture. It is necessary to renew a certain volume of H_2 lodged in the alternator by another volume coming from bottles H_2.

The principle of the interior cooling permitted the increase of the strength of the alternator and an efficient utilization of the hydrogen pressure (See Figure 10).

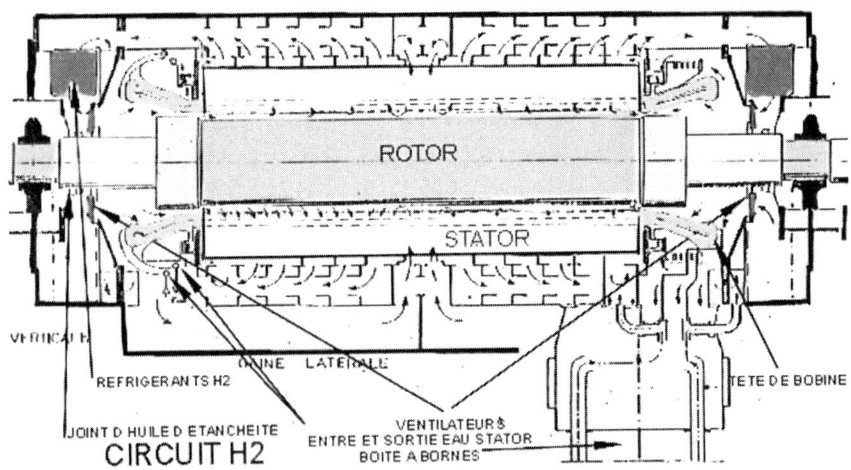

Figure 10. Cooling by hydrogen of an alternator.

Gas hydrogen is introduced in the alternator while manipulating the regulator of pressure or the regulator. When the purity of hydrogen measured is 95% or more on the meter, its feeding is stopped, the regulator of pressure is adjusted foreseen at the level and the pressure of the alternator is increased. Thus, gas hydrogen is introduced in the envelope, giving back the ready alternator to the working. The diagram of circuit of the system of gas control represents the position of every floodgate during the working.

Otherwise, the drier of gas, composed of a full reservoir of alumina activated (absorbing agent), of a heating device, of a puff, of a thermometer..., is installed between the circuit high pressure and the circuit bass pressure of the alternator so that gas crosses the drier all along the working of the alternator.

The turbo-alternator group of the center of Radès is cooled internally by gas hydrogen. As shows the diagram of circuit (See Figure 11), the device of gas control is composed by the following elements:

- A spray of carbon dioxide
- A device of gas hydrogen feeding
- A drier of gas
- A unit of surveillance of gas pressure / purity
- Command valves
- A meter of purity.

The pressure of gas hydrogen inside the alternator is maintained to a face value of 3 to ABS 6 bars, thanks to a regulator of pressure gone up on the collector of feeding in hydrogen.

In the same way, the purity of hydrogen in the alternator is always maintained to more of 95% and, when it descends to 90%, an alarm is given out, preventing the internal gas to compose an exploding mixture. The gas of carbon dioxide is used fluid how of sweep to fill to either hunt the hydrogen of the alternator in order to avoid that hydrogen and air won't be mixed in a critical condition.

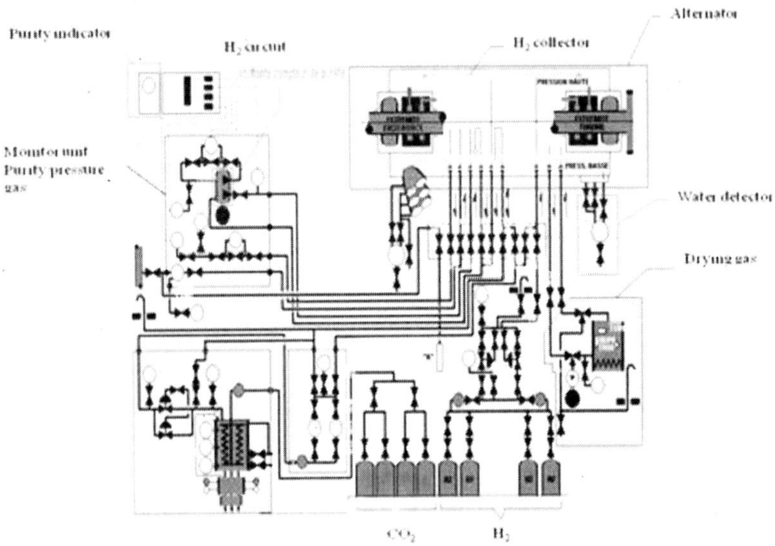

Figure 11. Circuit of the gas hydrogen of the TPP.

At the time of replenishment of the alternator by hydrogen, the dioxide of carbon is used to hunt the air of the alternator. The valve of safety is adjusted to an ABS 6 bar pressure so that, when an anomaly occurs in the circuit of gas of carbon dioxide, the pressure of the bottle is exercised on all tubings.

The CO_2 being heavier than air, it is provided in the alternator through the lower distribution hose. It is then necessary to measure the purity of gas to the top of the alternator: the lower hose leads to the valve of command in the post office of distribution that is opened opportunely and closed.

In the same way, the puff of the meter of the purity is starting up in the alternator with the spray in start. When the purity of carbon dioxide gotten on the meter of purity is besides 75%, the feeding is stopped and hydrogen is by following introduces to its room.

By reason of its relatively weak weight in relation to the CO_2, gas hydrogen is provided with the help of the superior hose of distribution of the alternator.

The purity of gas hydrogen must be measured to the bottom of the alternator that is for it the valve of command is opened appropriately and closed. A regulator of pressure is installed between the hose of gas feeding and the station of gas hydrogen in order to maintain the pressure of the internal gas to a value wanted of 1 to 8 Abs bars.

In this application, we present on the one hand, the programming of the general numbering, timetable and daily of the gas hydrogen consumption and on the other hand, the configuration of a new tabular circuit gas hydrogen containing the new information [21].

Figure 12 presents the new tabular of the Hydrogen circuit containing the new modifications.

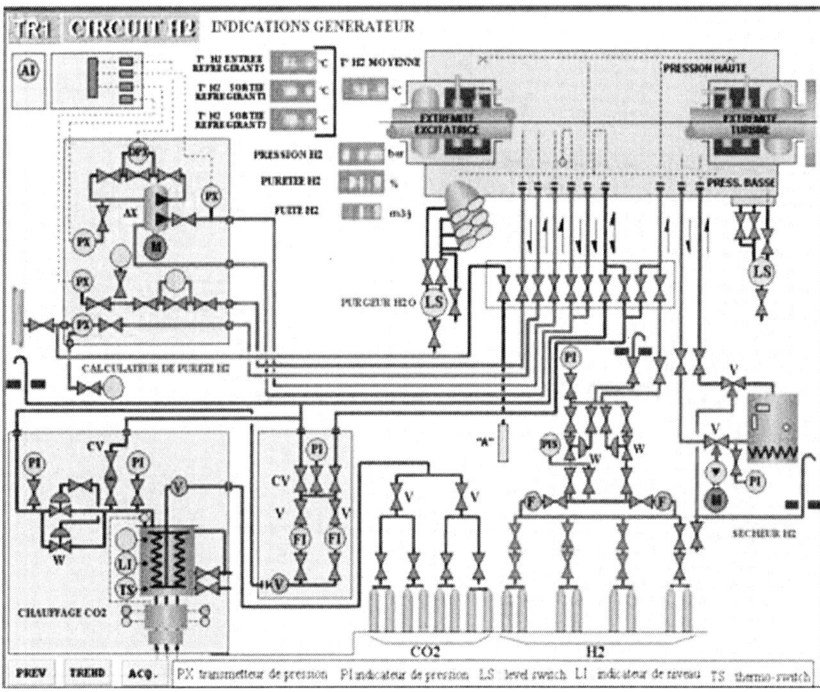

Figure 12. New tabular of the hydrogen circuit [21].

The TPP of Radès arranges a regulator of pressure that assures the feeding of H2. When the uncommunicative gas in the alternator and the pressure is increased the regulator in will be stopped. The calculation of hydrogen flights makes himself by hand therefore we cannot have an exact value on these flights. This value is not displayed on the SCADA system.

The daily taking of values of numberings of flights of hydrogen consumption decreases the precision of calculation of the output of every slice.

To remedy these problems we propose a solution of automatic reading of the value of the gas hydrogen flight. The proposed solution is to make the calculation of flights by the SCADA system and to program blocks of the daily calculation. This solution is automatic and cyclic where the period of the time is stationary.

The algorithmic of treatment is based on the concepts of block and diagram (or compound). Indeed, a block is a software entity that achieves a specific function more less complex (stake to the ladder, conversion, filtering, calculation, test of alarms, etc.) definite by its algorithm.

For the configuration stage, we used the ICC software (Integrated Control Configuration). This software permits the creation and configuration of the resident program in the CP60.

For the conception stage of the tabular, we used the FoxDraw software. This software possesses a library of components permitting to represent the various elements of an industrial installation.

5. APPLICATION OF THE SUPERVISION OF A WATER STEAM - CYCLE A TPP

Considering that the water used contains an elevated rate in dissolved salts and in matter suspended, it is indispensable to adopt a stage of pretreatment to assure the good working of the inverse osmosis installation and to protect modules against risks of usuries, corrosion and especially membrane calmative.

The pretreatment is constituted of two filtration chains each including a sand filter and an active coal filter. Thereafter, we present the two stations of the TPP: inverse osmosis and demineralization.

The control of the water quality is an important task to maintain the efficiency and the sure and continuous working of the power station. To guarantee the best water quality at the level of the water steam circuit, the TPP of Radès arranges an inverse osmosis station that permits to eliminate the majority of salts dissolved in the raw water before being treated in a demineralization station (SeeFigure 13). This stage serves to minimize risks of failing by corrosion of the turbine or the loss of the efficiency and the power.

The basic principle of the ion exchange consists in withdrawing ions (remaining salts that are lower to 8%) in solution in water is to recover an ion of value, either to eliminate a harmful or bothersome ion for the ulterior utilization of water.

The exchange of ions is a process which ions with a certain load contents in a solution are eliminated of this solution, and replaced in the same way by an equivalent quantity of other ions load gave out by the strong but the opposite load ions are not affected.

In the demineralization chain, osmosis water passes by the following stages:

- a weak cationic exchanger (CF1);
- a strong cationic exchanger (CF2);
- a weak anionic exchanger (AF1);
- a degasser;
- a strong anionic exchanger (AF2);
- a strong cationic exchanger (CF3);
- a strong anionic exchanger (AF3).

After the demineralization, the water must have a lower conductivity of 0.2 μS/cm, a pH between 6.5 and 7.5; silica < 30 ppb.

Figure 14 presents the cycle of the water treatment in the demineralization station.

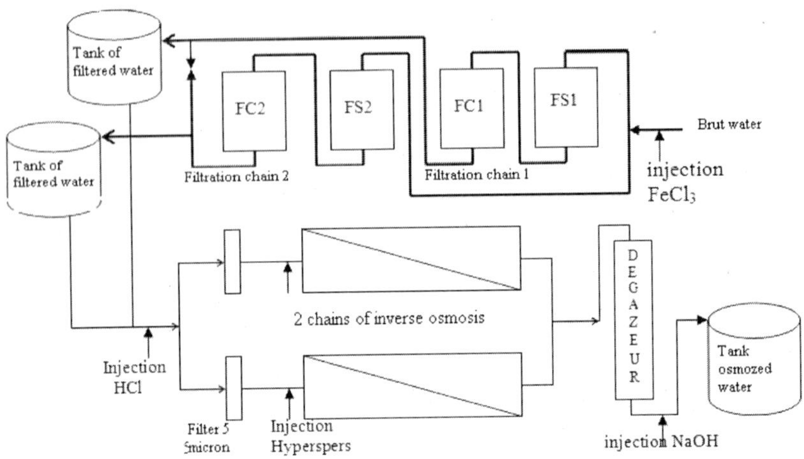

Figure 13. Legend:FS1: Sand filter of the filtration chain 1.FC1: Active coals filter of the filtration chain 1.FS2: Sand filter of the filtration chain 2.FC2: Active coals filter of the filtration chain 2.The bold lines present the water circuit in the two filtration chains and the light lines present the water circuit in the two inverse osmosis chains.

Functional diagram of the inverse osmosis station of the TPP.

In order to remedy to the absence of indication, of follow-up and of storage of the chemical characteristics of the water of the furnace, it is necessary to achieve an interfacing between the chemical sensors (pH meter and conductivity meter) and the stations of surveillance of the control room of the TPP.

The interfacing of the signals of the pH and the conductivity of the ball furnace is assured by a data configuration of both analogical and numeric signals and requires a unique code for every entrance which must be programmed in the data base system.

This application is declined in six stages:

Stage 1: Choosing the site of the signal (FBM module).

Stage 2: Programming both AIN and CIN blocks for the supervision of the signals pH (4 to 20 mA) and conductivity (alarm).

Stage 3: Testing both AIN and CIN blocks by injection of current and by short circuit.

Stage 4: Passing the cable between the sampling room and the SCADA room.

Stage 5: Connecting the signals in the two modules 10FBM215 and 10FBM325.

Stage 6: Conceiving a new tabular for the general vision of the sampling room.

The last stage of this application of interfacing consists in improving the tabular pH meter and conductivity meter.

Figure 15 presents the display of the sampling room containing the chemical analysis parameters of the water - steam cycle of the TPP.

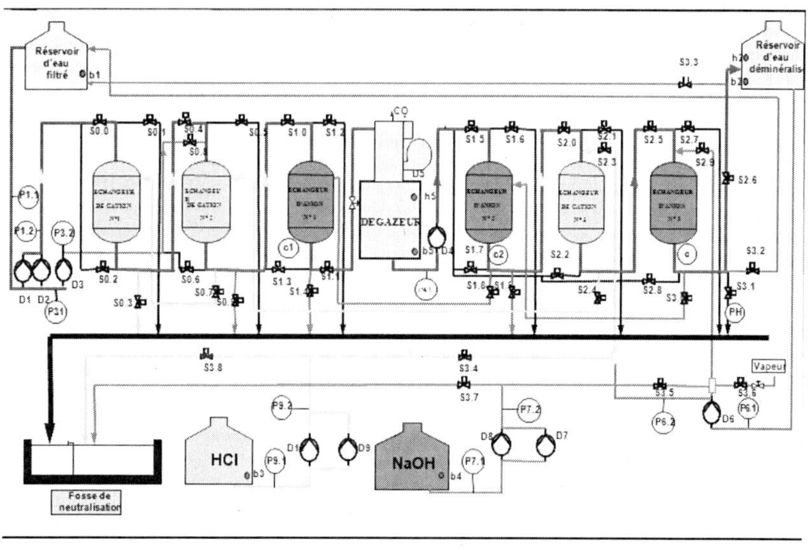

Figure 14. Demineralization station of the TPP.

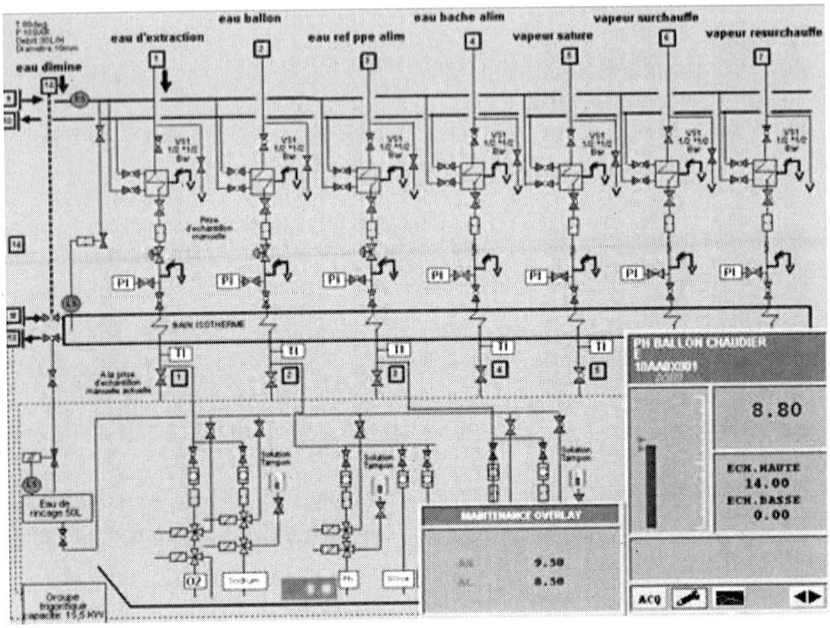

Figure 15. Display of the chemical analysis parameters [22].

6. APPLICATION OF THE SUPERVISION OF PUMPS VIBRATIONS OF A TPP

Systems of vibration surveillance are often equipped of measure chains for other complementary parameters, as the axial position, the crankiness, the differential dilation, the dynamic pressure, the speed of rotation and the temperature.

Among the new systems of measures, we mention notably IDS (system of icing detection) and AGMS (system of measure of the bore between the rotor and the stator) that complete a system of vibration surveillance efficiently, but that are also usable as of the autonomous specific systems [23].

The MMS system (System of Machine Surveillance) is the synthesis of the long experience of Vibro-Meter in the domain of the surveillance of machines and its expertise to master technologies of vanguard as for the manufacture of the electronic of surveillance.

The instrument of vibration control measures the vibration all the time when machines (turbine of power plant, big dimension compressor, pump, blower...) are in service. When the supervised vibration reached the amplitude of vibration, that is adjusted in advance, the instrument gives out an exit of point of alarm contact to give a warning to the working of the machine or gives out an instruction to stop the working of the machine, avoiding so the danger and accidents before they occur.

The mechanical vibration that is developed in a machine is controlled by a sensor of vibration and is converted in electric signal and this signal is introduced in an amplifier of vibration. In this amplifier, a signal that is proportional to the speed of vibration and supervised by an instrument of vibration control, and convert in a signal that is proportional to the displacement of the vibration, and this last is to its tower convert in a tension to continuous current, that is given back like signal to an indicator and a signal to the circuit of alarm.

The instrument of vibration measure used in our application is constituted by a sensor of vibration (Model U1-FH) and an instrument of vibration control (Model AVR-148). In fact, the sensor of vibration, model U1-F, is similar to the construction of a loudspeaker to permanent magnet. The sensor is attached to the machine on the one hand with screws and on the other hand to connect to the system of registration with the special cables.

With sensors of Vibro-Meter, we can measure in general most the critical parameters in the surveillance of machines, but particularly what concerns vibrations. In this domain, Vibro-Meter proposes a vast range of sensors, of conditioners of the signal as well as an effective signal transmission.

Sensors of proximity and other translators of displacement are based on currents of Foucault and present a high linearity with an active compensation of the temperature. Some sensors, as piezoelectric accelerometers, have favors to be deprived of the mobile pieces, what permits to guarantee their reliability and a long life span.

The piezoelectric sensor is used as detector of shock, vibration or percussion. It captures the mechanical vibrations that transmit itself in a material.

Figure 16 presents the diagram block of the system vibratory surveillance of a pump used in a TPP[24].

To achieve a complete monitor of surveillance, we always associate a module of treatment UVC 691 with a module of surveillance with a high performance PLD 772.

UVC 691 is a module of signal treatment assorted to different sensors and conditioners via a galvanic separation.

Most modules of Vibro-Meter provide unipolar signals in the range of 0 to 10 V DC. However, the PLD 772 can accept some bipolar signals in the range of 0 to ±10 V DC.

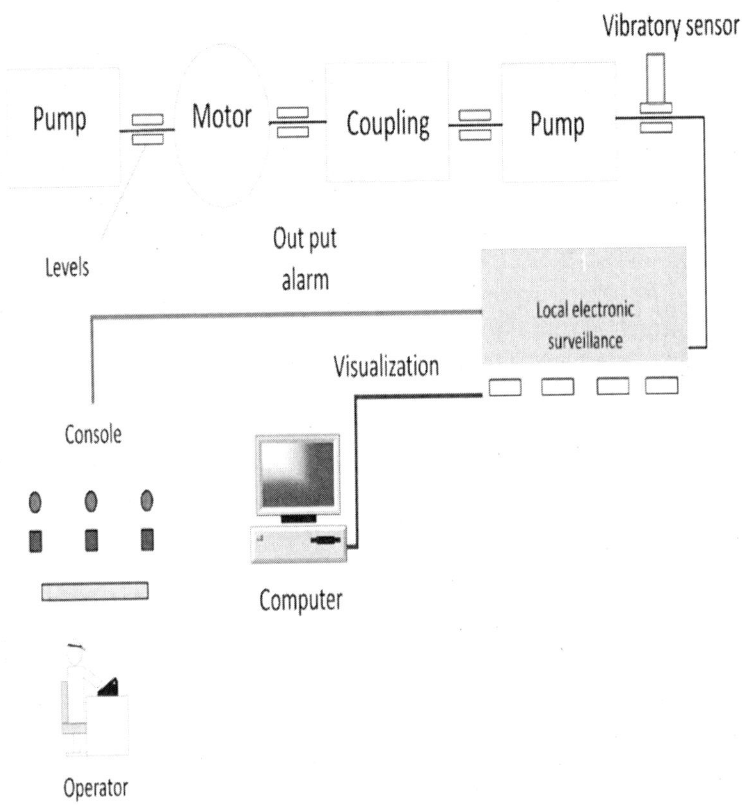

Figure 16. Block diagram of vibratory surveillance.

In fashion of programming of the PLD 772, the user has the possibility to define the calibration of the display and all parameters of alarm. While equipping the PLD 772 of an interfacing RS-485, the module is capable to the digital communication (Figure 17). Thus, a system of surveillance can make part of a cabled network. A computer detains the main computer role. All other modules PLDS 772 in racks are some secondary stations. Such a link between a system of surveillance and a main computer is in measure to do functions of programming from afar and of data transfer.

The central computer can read the calibration of every module at all times PLD 772. Instructions of set up and a special authorization permit to modify parameters of calibration or doorsteps of alarm of every surveillance module. Commands become thus easy and the result is from afar a fully programmable surveillance system.

The computer calls each module periodically to ask it the measured values (DC signals) and the state of alarm of every channel. Such a process of acquirement suits the registration of data and the creation of a data basis very well with the acquisition in DC in order to do an analysis of tendency subsequently using software of conditional maintenance.

We have built many displays with FoxDraw that are used by FoxView, and become the I/A Series interface to the pumping process of the TPP (See Figure 18).

Monitoring display and diagnosis display are important for the supervisory system. A global view of the system should appear the information needed for the pumping process [25].

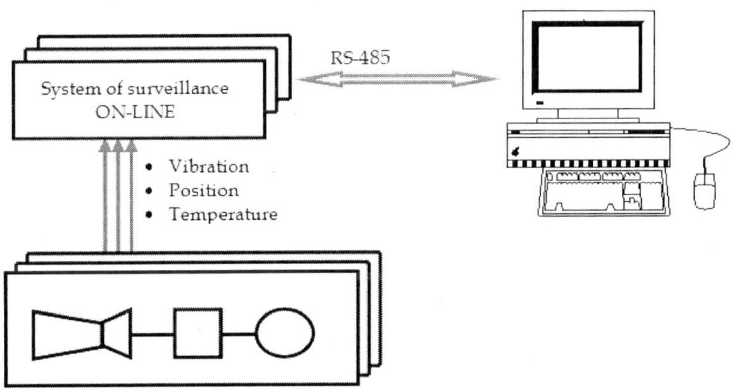

Figure 17. Transfer of data between units.

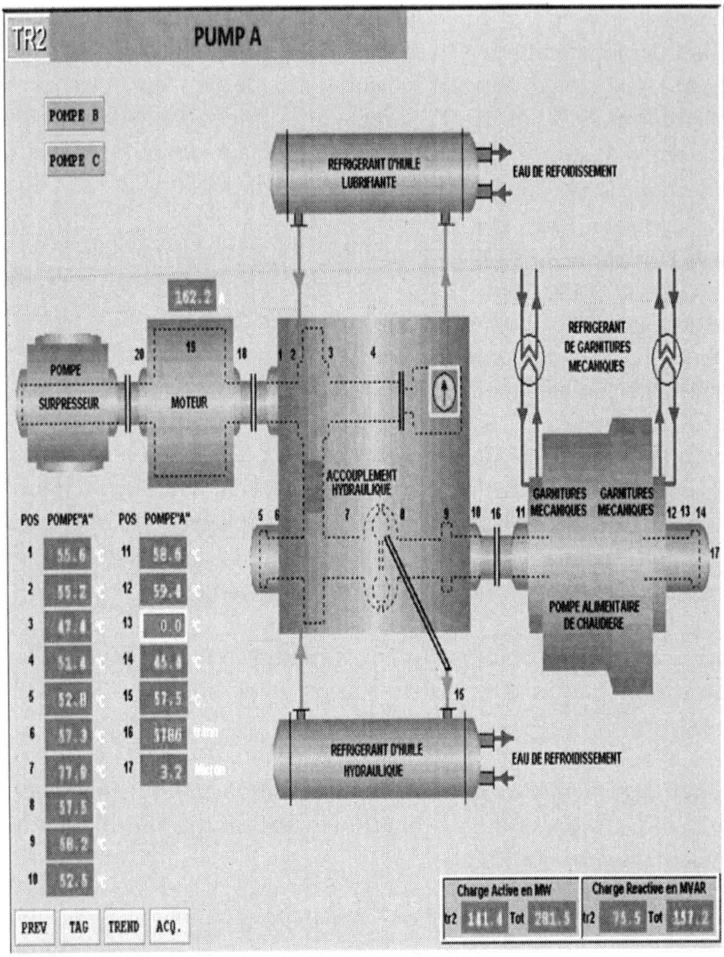

Figure 18. Display of the pump A of the TPP [24].

7. CONCLUSION

SCADA systems are used to control and monitor physical processes, examples of which are transmission of electricity, transportation of gas and oil in pipelines, water distribution, traffic lights, and other systems used as the basis of modern society.

SCADA system gathers information, transfers the information back to the central site, alerting the home station and carrying out necessary analysis and control, and displaying the information in a logical and organized fashion.

In this work, we presented on the one hand an application of Functional Analysis (FA) techniques on a SCADA system of thermal power plant (TPP) and on the other hand some examples of SACDA applications.

The first application of the SCADA system consists in integrating a module for the calculation of hydrogen flights in the alternator. The proposed solution was permit to elaborate a new tabular for the hydrogen circuit containing the new modifications. The second application of the SCADA system consists in interfacing the chemical analysis parameters of a water-steam cycle to the SCADA system. The third application is related to the supervision of a system of vibratory surveillance in a TPP. This application enables us the creating and the maintaining dynamically updating the pumping process displays.

This achieved applications is going to facilitate the work of both laboratory and instrumentation agents in the TPP.

Functional Analysis techniques seem to be a promising way because the major advantage of these kinds of techniques is due to the concept of function and abstraction hierarchy which are familiar to the human operator. These techniques permit the complexity of a system to be overcome.

In this paper, the application of SADT and SA-RT methods on a real system, a SCADA system of a TPP generates a source of useful information for the design of a supervisory system (monitoring and diagnosis displays, definition of alarms, etc.). So, research into the application of FA techniques for the design of a human centered supervisory system must be intensified in order to solve several difficulties and to improve their efficiency (tools to build the model, tools to check the validity of the model, etc.).

ACKNOWLEDGEMENTS

The author extends his especial thanks to Moncef Harrabi and Moez Lakhoua, Tunisian Society of Electricity and Gas (STEG) for their valuable help.

REFERENCES

1. D. Bailey, E. Wright, 2003 Practical SCADA for Industry, Elsevier

2. G. Clarke, D. Reynders, E. Wright, 2003Practical Modern SCADA Protocols, Elsevier,

3. E.I. Gergely, L. Coroiu, F. Popentiu-Vladicescu, 2010 Analysis of the Influence of the Programming Approach on the Response Time in PLC Control Programs, Journal of Computer Science and Control Systems, 31, Oradea, Romania, 61 64 .

4. S.A. Avlonitis, M. Pappas, , D. Avlonitis, K. Kouroumbas, N. Vlachakis, 2004 PC based SCADA system and additional safety measures for small desalination plants, Desalination 165, 165 176 .

5. G. T. Heng, 1996 Microcomputer-based remote terminal unit for a SCADA system, Microprocessors and Microsystems 20 1 39 45 .

6. J.H. Horng, 2002 SCADA system of DC motor with implementation of fuzzy logic controller on neural network, Advances in Engineering Software 33 361 364 .

7. M. Patel, G. R. Cole, T. L. Pryor, N. A. Wilmota, 2004Development of a novel SCADA system for laboratory testing, ISA Transactions 43 477 490 .

8. K. Munro, 2008SCADA- A critical situation, Network Security, 2008 1 , 4 6 .

9. J. Wiles, 2008Supervisory Control and Data Acquisition: Techno Security's Guide to Securing SCADA, 61 68 .

10. J. Wiles, 2008 Techno Security's Guide to Securing SCADA: A Comprehensive Handbook On Protecting The Critical Infrastructure, Elsevier

11. STEG 2008 Rapport Annuel de la STEG, Société Tunisienne de l'Electricité et du Gaz, .

12. Revue de l'électricité et du gaz 2008, Société Tunisienne de l'Electricité et du Gaz, 16.

13. K. Peffers, 2005 Planning for IS applications: a practical, information theoretical method and case study in mobile financial services, Information & Management, 42 3 483 501 .

14. DA. Marca, CL. McGowan, 1988 SADT: structured analysis and design technique, New York: McGraw-Hill Book Co., Inc.,.

15. K. Schoman, D.T. Ross, 1977 Structured analysis for requirements definition, IEEE Transaction on Software Engineering 31 1 6 15

16. M.N Lakhoua, 2009 Application of Functional Analysis on a SCADA system of a Thermal Power Plant, Advances in Electrical and Computer Engineering journal 9 2 90 98 .

17. M.N Lakhoua, 2009 Application of functional analysis for the design of supervisory systems: Case study of heavy fuel-oil tanks, International Transactions on Systems Science and Applications, 5 1, 21 33 .

18. H. Gomaa, 1994 Software design methods for the design of large-scale real-time systems, Journal of Systems and Software 25 2 127 146

19. P. Jaulent, 1992 Génie logiciel les méthodes : SADT, SA, E-A, SA-RT, SYS-P-O, OOD, HOOD, Armand Colin, Paris,.

20. M.N. Lakhoua, 2011 Application of SA-RT method to supervisory systems, Journal of Electrical Engineering, 1582-4594, Politechnica Publishing House, 11 2

21. M.N. Lakhoua, 2010 Application of a SCADA system on a hydrogen station, Journal of Electrical Engineering, 10 3

22. M.N. Lakhoua, 2011 “Systemic analysis of a supervisory control and data acquisition system”, Journal of Electrical Engineering 1582-4594, Politechnica Publishing House, 11 1

23. Système de surveillance des machines MMS 1990 , VIBRO-METER SA, Fribourg/Suisse,4.

24. M.N Lakhoua, 2010 Surveillance of pumps vibrations using a SCADA, Control Engineering and Applied Informatics, 12 1

25. M.N. Lakhoua, 2010 Application of a supervisory control and data acquisition system in a thermal power plant, symposium Large Scale Systems: Theory and Applications LSS' 10 12 11 14France.

Chapter 4

Feasibility of a Solar Thermal Power Plant in Pakistan

Ihsan Ullah[1], Mohammad G. Rasul[1], Ahmed Sohail[2], Majedul Islam[3] and Muhammad Ibrar[4]

[1] Central Queensland University, School of Engineering and Technology, Rockhampton, Queensland, Australia

[2] College of Electrical & Mechanical Engineering, Department of Mechanical Engineering, Rawalpindi, Pakistan

[3] Queensland University of Technology, Brisbane, Australia

[4] Islamia College University, Department of Physics, Peshawar, Pakistan

1. INTRODUCTION

Pakistan has been facing an unprecedented energy crisis since the last few years. The problem becomes more severe throughout the year. The current energy shortage crisis has badly hit Pakistan's economy where hundreds of industries have closed due to lack of electricity to fulfil their requirements. The energy supply and demand gap has risen to 5000 MW [1] and is expected to rise considerably in the coming years as shown in Figure 1. Table 1 shows the existing installed power generation in Pakistan.

Pakistan has a huge potential in renewable energy especially solar energy to fill this gap if utilized properly. Pakistan, being in the Sun Belt, is ideally located to take advantage of solar energy technologies. This energy source is widely distributed and abundantly available in the country. Pakistan receives 4.45- 5.83 kWh/m^2/day of global horizontal insolation as an annual mean value, with 5.30 kWh/m^2/day over most areas of the country [2, 3]. This minimum level of solar radiation (4.45 kWh/m^2/day) is higher than the world average of 3.61 kWh/m^2/day [4] which shows that Pakistan lies in an excellent solar belt range. Pakistan has six main insolation measuring stations, namely Karachi, Islamabad, Lahore, Quetta, Peshawar and Multan and 37 observatories distributed fairly well over the entire country, recording sunshine hours as shown in Table A in Appendix. From the sunshine hours data it can be seen that most of the cities mentioned receive more than 250 sunshine hours a month.

2006 energy policy has resulted in few practical steps taken for utilizing the abundantly available solar resource in Pakistan. A Solar Water Heating System has been installed in a Leather Industry for first time in Pakistan. The system, using 400 m^2 Evacuated Collector tubes, provides heated water at 70 to 80 °C (at least 10 degree rise to the incoming water) to the already used boiler system, thus saving 33% of the cost. The Project was funded by Higher Education Commission (HEC) under University-Industry Technological Support Program (UITSP).

Solar water heating technology is relatively mature technology in Pakistan but its higher capital cost compared to conventional gas heaters has limited its use so far. This technology is widely recommended by a number of public sector organizations in northern mountains where natural gas is limited and difficult to supply. The solar water heaters are now being commercially produced in the private sector.

More than 2000 low cost solar cookers are used in Pakistan for cooking purposes. Similarly, solar dryers are used in Gilgit and Skardu (Northern part of Pakistan) to dry large quantities of fruits such as apricot and transport and sell them later in the urban areas, thus bringing economic prosperity to the area.

Fresh water unavailability in large parts of Baluchistan, Sind and southern Punjab is a critical issue. Two solar desalination plants consisting of 240 sills each with a capacity of 6000 gallons of seawater/day have been operational at Gawadar in Baluchistan province. A number of such schemes are under active consideration by local governments in Baluchistan and Thar [5].

The need for constructing solar power plants has been realized both at federal and provincial governments. The government of Sindh recently signed a Memorandum of Understanding with German company Azur Solar for building a 50 MW solar power plant at Dhabeji in District Thatta. The firm Azur Solar will initially set up a 60 kW solar power station to provide free electricity to backward 'goths' (Villages), schools and basic health centres of Badin.

Both, technical and economical feasibility of a 50 MW solar thermal power plant using Stirling-Dish (SD) technology in Jacaobabad in Sindh Province of Pakistan is analysed in this chapter. The performance and environmental aspects of Stirling dish for power generation with and without solar energy is examined, discussed and compared. The solar data was collected from Bureau of Meteorology (BoM) of Pakistan and Life cycle cost analysis is performed to determine the economic feasibility of the solar thermal power plant. This study reveals that Jacobabad falls within the high solar energy belt and has minimum radiation energy of 4.45 kWh/m^2/day (which is higher than the world average of 3.61 kWh/m^2/day) to produce electricity, even during the low sunshine and cloudy days. The study concluded that solar thermal power plant in Pakistan is feasible using solar Stirling dish technology.

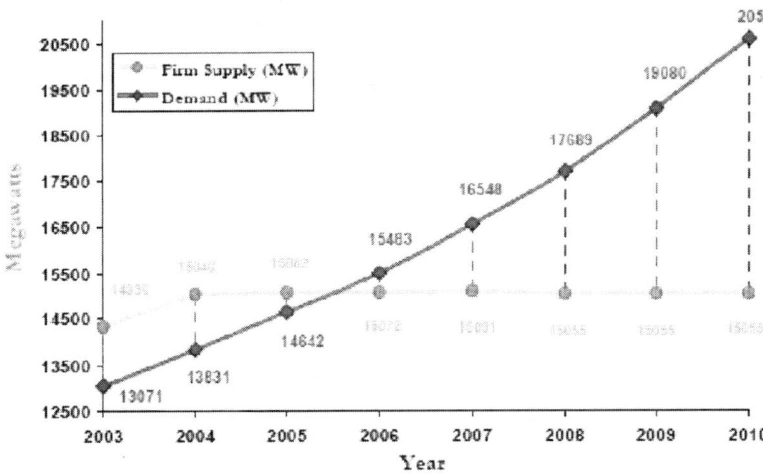

Figure 1. Generation capacity and demand forecast [6]

Table 1. Existing installed generation capacity in Pakistan [1]

Type of Generation	Installed Capacity (MW)	Derated / Dependable Capacity (MW)	Availability (MW)	
			Summer	Winter
WAPDA Hydro	6444	6444	6250	2300
GENCOs	4829	3580	2780	3150
IPPs	6609	6028	5122	5402
Rental	285	264	250	250
Total	18167	16316	14402	11102

2. SOLAR STIRLING ENGINES

Solar thermal electric power generating systems have three different design alternatives, as follows;

- Power tower: a solar furnace using a tower to receive the focused sunlight
- Parabolic trough collector: focus systems that concentrate sun rays onto tubes located along the focal line of a parabolic shaped trough.
- Parabolic dishes: focus systems where sun light is reflected into a receiver at the dish's focus point [7, 8].

High optical efficiency and low start-up losses make dish/engine systems the most efficient (29.4% record solar to electricity conversion) of all solar technologies [9]. In addition, the modular design of dish/engine systems make them a good match for both remote power needs in the kilowatt range as well as hybrid end-of-the-line grid-connected utility applications in the megawatt range as shown inFigure 2 [9].

Solar Stirling engines can be classified into two categories;

- Free Piston Stirling engines: are those which have only two moving parts i.e. the power piston and the piston, which moves back and forth between springs. A linear alternator extracts power from the engine through power piston. Electricity is produced internally and therefore, there is no need for sliding seal at the high pressure region of the engine and no lubrication is required too [10].
- Kinematic Stirling engines: are those in which both the power piston and displacer (expansion and compression pistons) are kinematically (mechanically) connected to a rotating power output shaft.

Kinematic engines work with hydrogen as a working fluid and have higher efficiencies than free piston engines. Kinematic sterling engines have sealing problems and complicated power modulation. Sealing problems can be avoided by integrating a rotating alternator into the crankcase [10]. The power modulation can be sorted out by; (a) Varying the piston stroke (b) varying the pressure level of the working space [10].

Free piston engines have simple design as there is no connection between power piston and displacer. There is no need for working fluid make-up system being hermetically sealed as is required in the case of kinematic Stirling engines [11]. Free piston engines work with helium and do not produce friction during operation, which enables a reduction in required maintenance.

The solar Stirling engine is environment friendly as the heat energy comes from the sun and therefore almost zero emission. Similarly emissions from hydrocarbons combustion are very low as the fuel is burnt continuously at almost atmospheric pressure compared to the interrupted combustion in diesel and petrol engines. The quantities of CO produced and of unburnt hydrocarbons HCs are very low due to (i) the combustion of fuel in a Stirling engine occurs in a space surrounded by hot walls under adiabatic conditions and (ii) the latitude in the choice of air to fuel ratio. But unfortunately, the more efficient combustion of a Stirling engine results in relatively more CO_2 produced compared to an equivalent internal engine. Similarly the formation of NO_x are lower due to the short residence time of the gases at the high temperature, lower peak temperatures than in internal combustion engine and the continuous combustion. The emission of NO_x can be further reduced: (a) by recirculation of part of the flue gases along with incoming combustion air and (b) lowering the flame temperature. Stirling engine is intrinsically cleaner than all current heat engines in terms of emission of toxic or other polluting substances [11].

Figure 2. General description of Stirling EuroDish system [12]

3. STIRLING-DISH SOLAR ELECTRIC POWER GENERATING SYSTEM

Stirling-Dish (SD) systems are small power generation sets which generate electricity by using direct solar radiation. The capacity of a single unit is typically between 5 and 25 (50) kW_{el}. This size and the modularity of the single units qualify the Dish-Stirling system for very flexible applications. They are ideal for stand-alone or other decentralised applications [13]. The size of the solar collector for Stirling Dish is determined by the desired power output at maximum insolation levels (1 kW/m^2) and the collector and power conversion efficiencies. A 5 kW Stirling Dish system requires a dish of ca 5.5 m, in diameter, and a 25 kW system requires a dish of ca 10 m in diameter [10]. The schematic diagram of solar Stirling dish system is shown in Figure 3. The parabolic concentrator reflects the solar radiation onto a cavity receiver which is located at the concentrator's focal point. The heat exchanger (receiver) absorbs the solar radiation and thus heats the working gas (Helium or H_2) of the Stirling engine to temperatures of about 650° C. This heat is converted into mechanical energy by the Stirling engine. An electrical generator, directly connected to the crankshaft of the engine, converts the mechanical energy into electricity (AC). A sun tracking system rotates the solar concentrator continuously about two axes to follow the daily path of the sun to constantly keep the reflected radiation at the focal point during the day [13].

4. RESULTS AND DISCUSSION

4.1. Technical Feasibility

A SD solar electricity system of 25 kW with design characteristics shown in Table 2 was used for analysis in this study [10]. Stirling Dish has the ability to operate earlier and later every day and can also be functional in cloudy conditions when solar energy is < 2 kWh/m^2. Another advantage of SD is it can generate power between passing clouds due its low thermal inertia. Jacobabad's Latitude is 28°18′ N, Longitude is 68°18′ E and elevation is 55m. Jacobabad is among the hottest and the most arid areas of Pakistan, having weather conditions quite similar to those of a desert where direct normal insolation is high (1735 kWh/m^2/year) [15] as shown in Table 5 coupled with low land cost, which is really cheap compared to international costs of land in Europe or USA.

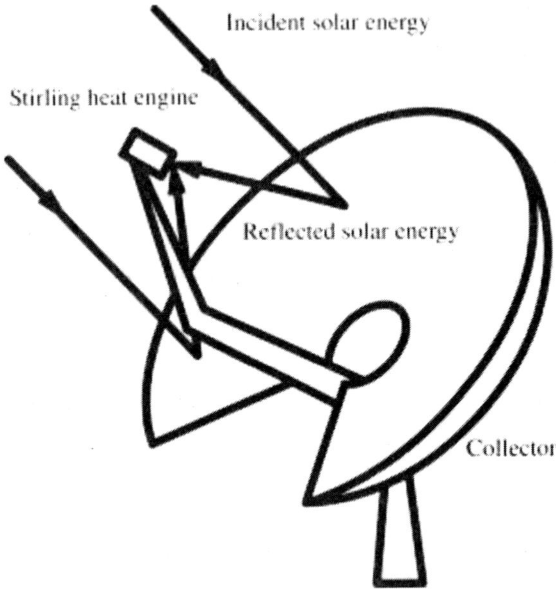

Figure 3. Schematic diagram of solar Stirling dish system [14]

The Jacobabad's 10 years sunshine hours data was provided by Pakistan Meteorological Department as shown in Table 4. It can be seen from Table 4 that there are very good sunshine hours throughout the year and the average sunshine hours is 9 for a month. It is slightly less than that of National Renewable Energy Lab NREL total sunshine hours [17] but even then it is highly suitable for a solar power plant.

It is the hottest location in Pakistan as the temperature is around 45 to 50°C in summer and10 to 25°C in winter as shown in Figure 4. Temperature data given in Table 3 also shows a low temperature mean value of 19.95°C, high temperature mean value of 34.15°C and Temperature mean value of 27.05°C very good hot conditions at Jacobabad [16], which are highly conducive for a solar thermal power plant.

Direct normal insolation data from National Renewable Energy Lab (NREL) USA, shown in Figure 5and Table 5, was compared with the data from BoM (Pakistan) for assessment of the Jacobabad site. It can be seen that there is not a huge difference between the NREL data and the data received from Pakistan Bureau of Meteorology.

By comparing the sun shine hours of these cities available data, it was seen that sun shine hours in Jacobabad are more than other cities as shown in Table A1 of Appendix.

4.2. Economical Feasibility

The selected economic indicators for this analysis are: Levelized cost of electricity, net present value (NPV) and total payback period (Tpb) [14].

Table 2. Design characteristics of 25 kW solar Stirling Dish [12]

Design Characteristics of a solar electricity system SD 25 kW	
Concentrator	
Glass area	91.01 m^2
Aperture area	87.67
Focal length	7.45
Glass type	No. 82 commercial grade float. Thickness 0.7mm
Radius of curvature	599. 616, 667, 698"
Waviness	<0.6mr
Reflectivity	>90%
Module dimensions	11.89 mH, 11.28 W
Module weight	6.934 kg
Stirling engine (kinematic)	
Engine dry weight	225 kg
Displacement	380 cc
Engine dimensions	66 cm W, 71 cm H, 58 cm L
Number of pistons	4 double acting
working fluid	H_2 or He
working fluid pressure	20 MPa
Operating temperature	720°C
Power control	Fluid pressure
Cooling	Water/forced air fan
Output power	27 kW (max), 22 kW (rated)
Rated power efficiency	38-40%
Power conversion unit	
Weight	>680 kg
Alternator	Induction, 1800 rpm
Alternator efficiency	92-94 %
Electrical power	480 V, 60 Hz, three phase
Gross power rating	25 kW at 1000 W/m^2
Peak net power efficiency	29-30%
Minimum insolation	250-300 W/m^2
Dimensions	W=168 cm, H=122 cm, L=183 cm

Table 3. Climate average weather data for Jacobabad [16]

NOAA Code	Statistics	Units	Jan	Feb	Mar	Apr	May	Jun	Jul	Aug	Sep	Oct	Nov	Dec	Average
0101	Temperature Mean Value	F	59.2	64.2	74.8	86.4	94.8	98.4	94.8	91.8	88.5	82	71.8	61.5	80.7
0109	High Temperature Mean Daily Value	F	72.7	77.4	88.3	100	110	112	105	101	98.6	95.5	86.2	75.4	93.5
0110	Low Temperature Mean Daily Value	F	45.9	50.9	61.3	72.1	80.1	84.9	84.6	82.9	78.5	68.5	57.4	47.7	76.9
0615	Precipitation Mean Monthly Value	Inches	0.1	0.3	0.4	0.1	0.1	0.2	1.5	1.1	0.5	0.1	0	0.2	0.4
0101	Temperature Mean Value	C	15.1	17.9	23.8	30.2	34.9	36.9	34.9	33.2	31.4	27.8	22.1	16.4	27.05
0109	High Temperature Mean Daily Value	C	22.6	25.2	31.3	38	43.1	44.3	40.6	38.2	37	35.3	30.1	24.1	34.15
0110	Low Temperature Mean Daily Value	C	7.7	7.7	16.4	22.3	26.7	29.4	29.2	28.3	25.9	20.3	14.1	8.7	19.95
0615	Precipitation Mean Monthly Value	mm	3.1	7.1	10.3	2	1.7	4.7	36.8	26.3	11.2	2.3	1.2	3.7	9.2

$$C_L = \frac{I+O}{E_1 \sum_{t=1}^{n} \frac{1}{(1+k)}} [\$/kWh] \tag{1}$$

where I, O E_1, k and n are values of investment, operations and maintenance, energy produced by the plant in the first year (taken as the average annual energy produced), discount rate and number of years envisaged as the plant life time. Therefore, levelized cost is the total cash flows of a plant divided by the energy produced over the lifetime of a plant. It has been assumed that land is provided by the government. The operation and maintenance costs (O&M) cost in Pakistan for a solar thermal power plant is $ 0.001/kWh [18]. The energy inflation is considered to be negligible. We assume 60% sunshine hours are available throughout the year for electricity production to account for cloudy conditions and other miscellaneous factors. We shall use a discount rate of 10% for this analysis. The assumption and data used for the analysis is shown in Table 6, a [19]

Table 4. Jacobabad sunshine hour's data [3]

JACOBABAD SUNSHINE HOURS PER MONTH												
Year	Jan	Feb	Mar	Apr	May	Jun	Jul	Aug	Sep	Oct	Nov	Dec
2000	213.3	248.1	240.8	273.7	364.1	361.2	330.0	307.4	297.8	308.5	241.6	245.2
2001	269.7	207.8	266.0	261.6	342.1	303.1	209.2	315.4	290.3	287.2	287.4	249.6
2002	238.5	207.4	242.0	263.2	330.7	342.4	300.6	240.6	281.7	300.1	243.3	239.9
2003	200.8	194.0	222.4	128.3	298.2	320.0	228.6	270.1	211.5	294.7	254.2	233.5
2004	200.8	194.0	255.0	128.3	262.2	275.6	270.3	256.3	237.5	318.2	294.5	238.1
2005	200.8	194.0	218.7	128.3	298.5	328.6	341.9	333.4	284.5	313.3	271.7	264.4
2006	216.0	193.9	251.8	247.4	341.9	331.4	312.7	251.2	307.7	310.9	221.4	203.2
2007	213.4	200.0	293.0	319.0	385.5	331.5	313.0	341.5	343.5	335.0	289.5	270.5
2008	171.0	254.5	279.0	273.5	359.0	288.0	323.0	341.5	328.0	315.5	289.0	258.5
2009	164.7	255.9	289.1	320.6	252.0	252.0	269.0	372.0	281.2	309.5	285.3	247.4

Table 5. Average Direct Normal Insolation data for Jacobabad [**15**]

Month	Monthly DNI (kWh/m²)	Average Day Length (Hours)
January	114.42	10.4
February	127.37	11
March	149.65	11.8
April	165.61	12.7
May	169.83	13.4
June	177.66	13.8
July	168.32	13.6
August	131.59	13
September	157.78	12.2
October	154.77	11.3
November	115.63	10.6
December	102.98	10.2
Monthly Average	144.63	12
Annual Total	1735.56	

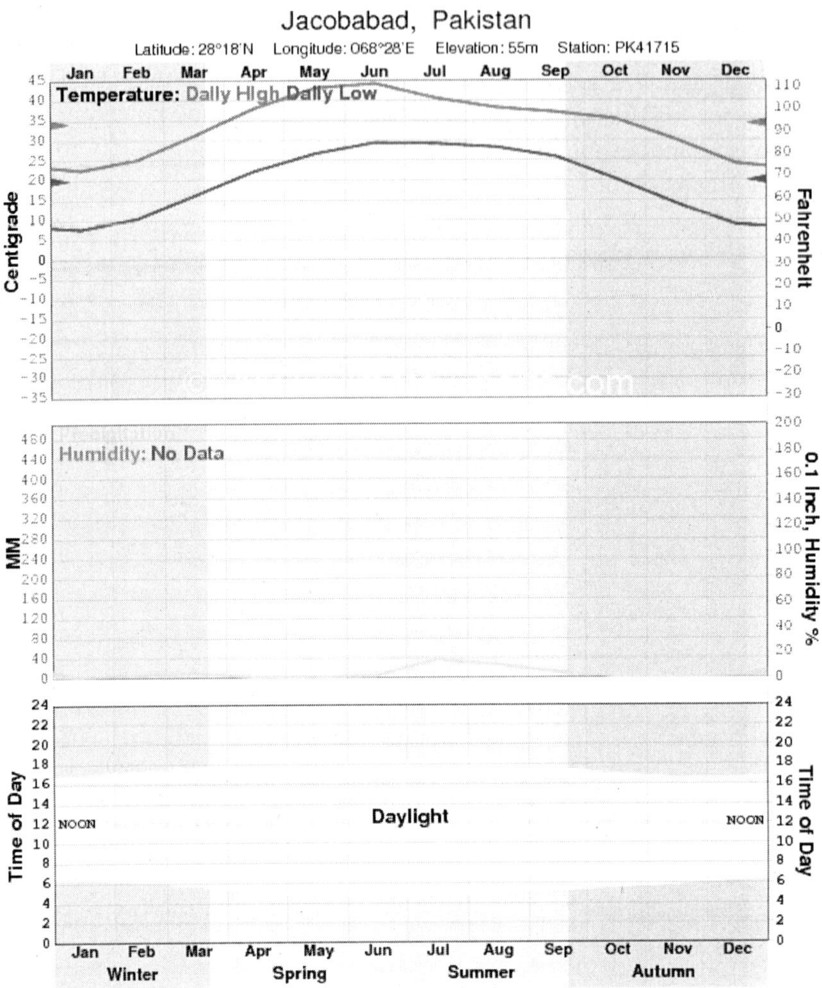

Figure 4. Jacobabad daily temperature – daylight chart [16]

Net present value (NPV) which represents the discounted cash flow over the lifetime of a plant can in general be stated as

$$NPV = (1\text{-}T)\ (E - O) - I \tag{2}$$

Where T is marginal tax rate, E is income of the plant, O is O&M costs and I is initial investment.

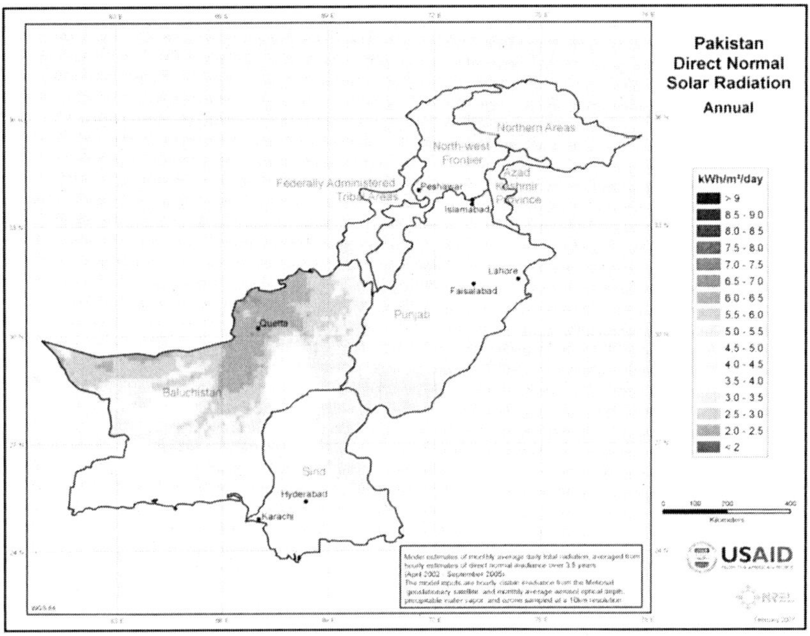

Figure 5. Pakistan annual direct normal solar radiation [17]

Table 6. Assumptions and data

Assumptions and data used	
Technical data	
Total power (MW)	50
Annual solar radiation (kWh/m²)	1735
Annual generated electricity (MWh)	18970
Discount rate	10%
Lifetime (years)	25
System purchase price (US$/kWh)	4540[a]
Fixed cost	
Procurement of equipment (M$)	227
Transport & Installation (M$)	3[a]
Other costs (M$)	3[a]

Total payback period is the ratio of the initial capital investment I to the net income (difference between the value of the energy produced in the first year of operation, E_1, and the sum of the expenditure on operation and maintenance in the first year, O_1,

$$T_{pb} = I/(E_1 - O_1) \tag{3}$$

Table 7. Summary of calculated quantities for a 50 MW solar thermal power plant

Quantity	Calculated Value
Value of investment	233 million US $
O&M Costs	0.475 million US$
Levelized electricity cost	24.4 cents/kWh
Sale price	60 cents/kWh
Net present value	22 million US$
Payback Period	23 years

Solar dish technology can be a promising technology which can be deployed in Pakistan in small scale projects producing 25 kW electricity for remote areas, especially, in Sindh and Baluchistan and Federally Administered Tribal Areas (FATA), where there is enough solar potential for producing electricity through dish technology. Jacobabad has a great potential for a solar power plant based Stirling dish. The solar insolation is in the required limits which is more than 1700 kWh/m^2/year coupled with prolonged sun shine hours in those areas make this technology a viable option for producing electricity. Cost analysis shows that this technology is viable although it is expensive at the moment compared to fossil fuels based electricity. The prices will become competitive if the government of Pakistan withdraws the subsidies on fossil fuels and allocate it to solar based electricity. The calculations show that cost of electricity generation is higher than the existing water, natural gas and fossil fuels based generation which is 3 to 12 USD cent/kW [20] and therefore, government subsidy is necessary to make it competitive in the market. If government can provide subsidy on renewable power then it will attract more investment both at local and international level.

5. CONCLUSIONS

Pakistan has a huge potential for solar energy potential especially very high in Quetta and Karachi where a solar thermal power plant is viable. Government of Pakistan has to take strong policy and marketability measures that will establish and strengthen this environment friendly technology in the country. Solar power has very little impact on environment. This makes it one of the cleanest sources of power generation available to mankind. An operating solar power plant produces no air and noise pollution. Furthermore there is no hazardous waste produced in the production of electricity and it also does not require and transportable fuel. The use of solar electric systems is also known to reduce local air pollution. This results in the reduction in the use of kerosene and other fuels for lighting purposes. Solar power systems also help in the abatement of CO_2 gases.

Jacobabad is one the hottest place in the country and therefore, has the potential to have solar power plants installed there. Analysis for Stirling dish shows that it is feasible to install such systems there as the insolation is suitable for a solar power plant. The sunshine hours are also good coupled with good infrastructure i.e. transmission lines, natural gas and coal are in close proximity and easy access to national highway makes it an ideal place for a solar thermal power plant. Gawadar and Karachi ports are in the range of 500 to 700 km and also 132/220 kV transmission lines pass through the area for grid connectivity of the power produced by this plant. A positive net present value coupled with a reasonable payback period of 23 years indicates that this plant is a good option.

REFERENCES

1. Water & Power Development Authority of Pakistanhttp://www.wapda.gov.pk/htmls/power-index.html,June 09, 2012

2. Alternative Energy Board (AEDB)Power sector situation in Pakistan, Alternative Energy Board (AEDB), Deutsche Gesellscharft fur Technische Zusammenarbeit (GTZ) GmbH, September 2005

3. Pakistan Meteorological DepartmentGovernment of Pakistan, May 2010

4. I. A Raja, and J. W Twidell, Statistical analysis of measured global insolation data for Pakistan, Renewable energy: 421992161994

5. World Energy CouncilRenewable energy in South Asia: Status and prospects. London, UK; November 2000

6. Private Power Infrastructure BoardMinistry of Water & Power Govt. of Pakistan

7. H Klaiss, R Kohne, J Nitsch, U Sprengel, Solar thermal power plants for solar countries- technology, economics and market potential. Appl Energy 19955216583

8. K Marketaki, V Gekas, Use of the thermodynamic cycle Stirling for electricity production. In: Proceedings of the 6th Panhellenic Symposium of Soft Energy Sources, 1999283290

9. Overview of solar thermal technologies by Centre for renewable energy resourcesUSA, May 1997

10. Teocharis TsoutsosVasilis Gekas, Katerina Marketabi, Technical and economical evaluation of solar thermal power generation, Renewable Energy 28 (2003873886

11. C. M Hargreaves, The Philips Stirling engine. Amsterdam: Elsevier Science, 1991

12. P. J De Graaf, Multicylinder free-piston Stirling engine for application in Stirling-electric drive systems. In: Proceedings of the 26th Intersociety Energy Conversion Engineering Conference- IECEC '91. Boston, MA, USA, 19915205210

13. Schlaich Bergermann und Partner GbREuroDish-Stirling System Description, Schlaich Bergermann und Partner GbR, Structural Consulting Engineers, Stuttgart June 2001

14. Li YaqiHe Yaling, and Wang Weiwei, Optimization of solar-powered Stirling heat engine with finite-time thermodynamics, Renewable energy 36 (2011421427

15. Waqas NasrUllah Khan ShinwariFahd Ali, and A.H. Nayyar, Electric power generation from solar photovoltaic technology: Is it marketable

in Pakistan? The Pakistan Development Review 43: 3(Autumn 2004267294

16. ClimateGlobal Warming, and Daylight Charts and Data; http://www.climate-charts.com/Locations/PK41715php,August 2010

17. National Renewable Energy Lab; http://wwwnrel.gov/analysis/tech_cap_factor.html, National Renewable Energy Lab USA, September 2010

18. Policy Framework and Package of Incentives for Private Sector Power Generation Projects in PakistanPrivate Power Cell, Ministry of Water and Power, Government of Pakistan, 1994Annexure I, 1

19. In the fieldEPRI Estimates Performance, Costs for Solar Thermal Plants,http://mydocs.epri.com/docs/CorporateDocuments/EPRI_Journal/2012Spring/1025049_InTheField.pdf,EPRI Journal Spring 2012

20. National Electric Power Regulatory AuthorityGovernment of Pakistan, http://nepra.org.pk/tariff_ipps.htm,November 2012

Chapter 5

Green Electricity from Rice Husk: A Model for Bangladesh

A.K.M. Sadrul Islam[1] and Md. Ahiduzzaman[2]

[1] Department of Mechanical and Chemical Engineering, Islamic University of Technology (IUT), Gazipur, Bangladesh

[2] Department of Agro-processing, Bangabandhu Sheikh Mujibur Rahman Agricultural University, Gazipur, Bangladesh

1. INTRODUCTION

Bangladesh has reportedly over one hundred thousand Rice Mills - large, medium and small. They process paddy, using mostly the parboiling practice. The rice mills use the thermal energy of steam generated in boilers, which are fired by rice husk - a byproduct of paddy processing in the rice mills and which is globally well-known as very convenient source of dry biomass energy of reasonable heat value. In global context, annual rice husk production is 137 million tonnes whereas, in Bangladesh, about 9.0 million tonnes of rice husk is produced reported in 2011 [1]. In Bangladesh context, out of 37.08 million tonnes of total biomass produced from agro-residues, rice husk contributes about 26% by mass [2]. At present about 67-70% of rice husk is consumed for steam producing in rice mills [3,4]. With a few exceptions, presently most of the 'boilers', used in these rice mills of Bangladesh are very inefficient. This results in a huge wastage of rice husk, which is an important source of biomass. Preliminary estimates are indicative that at least 50% of the rice husk produced could be saved and made surplus for its better use as input for small power generation [5]. Energy demand for rice processing is increasing every year due to the increased production of rice for ensuring food security of the population. More energy efficient rice parboiling boilers is to be suggested to replace the existing inefficient rice parboiling systems. So that the surplus amount of rice husks could be available for other sectors. If this husk is to be briquetted, then it will be an alternative fuel for replacing wood-fuel. Rice husk based co-generation system could be an alternative for large rice mills, or even for a cluster of small to medium rice mills. If the electricity were generated from rice husk then net carbon dioxide emissions could be reduced [6]. As a matter of basic dissemination strategy, applicable to most of the developing countries, initially the entrepreneurs/investors feel shy to invest in a relatively new area technology, as they lack the confidence on the technical functioning and financial/economic viability of ventures in such an area. Before launching a husk based power generation technology, it is important to know the production and distribution of available rice husk resource at different location. There are some different options to enhance the availability of rice husk such as increasing the un-parboiled process, improvement of thermal and combustion efficiency of existing husk fired boiler for parboiled process of paddy. Another important issue is the sustainable supply of rice husk in future for husk based power generation plant. Therefore, rice husk production and supply system is analyzed for electricity generation from husk in Bangladesh.

2. METHODOLOGY

There are main two sector of rice process viz. household level process and processed in rice mill. Husk obtained from household level process cannot be collected due to its wide spread localized. Therefore, husk from household level process is not considered as supply source of husk for electricity generation. Paddy processed in rice mill is considered the source of husk supply. There are several rice processing zones in the country. To conduct the investigation four major rice processing zones were selected to estimate potential husk available for electricity generation. The selected rice mill clusters are located at Dinajpur, Naogaon, Bogra, and Ishawrdi (Pabna). A survey questionnaire was used to collect the information on monthly quantity of paddy processed, amount of husk produced, amount of husk consumed for rice processing purposes and the surplus amount of rice husk (Survey questionnaire is shown in Appendix). Two hundred number of rice mills were surveyed in this study.

Another study is conducted to estimate the total electricity generation from rice husk in Bangladesh. LEAP (Long-range Energy Alternative Planning System) tool is used to make three different scenarios for consumption of energy. The basic parameter for estimating scenarios data are shown in Table 1.

Table 1. Assumption of activity levels at different scenario

Activity level	Reference scenario	Scenario-1
Paddy processed in mill	70%	90%
Parboiled processed paddy	90%	90%
Unparboiled processed paddy	10%	10%
Traditional parboiling	100%	5%
Improved parboiling	0	95%

3. RESULTS AND DISCUSSION

3.1. Available Husk At Rice Mills

Annual paddy processing capacity and surplus amount of husk from the four selected rice processing zones are analyzed. The annual paddy processing capacities were estimated to be 936923, 394976, 1366980 and 925048 tonne per year for Pulhat (Dinajpur), Bogra, Naogaon and Ishawrdi cluster, respectively. Total paddy processed in these clusters was 3.62 million tonne. Amount of surplus husk for selected rice processing zones were estimated to be 82904, 586925, 192551 and 121209 tonne per year for Pulhat, Bogra, Naogaon and Ishawrdi cluster, respectively. Total surplus of husk were 455356 tonne per year in the selected four clusters.

3.2. Electricity Generation Capacity Based On Rice Husk Energy At The Study Areas

Electricity generation from rice husk depends on the availability of raw material and the technology for conversion rice husk to energy. For a steam turbine power plant consumption of rice husk is 1.3 kg per kWh electricity as reported by Singh [7]. On the other hand for a gasification power plant consumption of rice husk is 1.86 kg per kWh electricity generation [8]. Based on this assumption the capacity of electricity power generation was estimated for the study areas (Table 2). The potential power capacities is estimated to be 41450 kW in four selected zones considering the steam turbine technology, whereas the corresponding power capacities is estimated to be 29050 kW considering gasification technology.

Table 2. Summary of electricity production from rice husk

Rice processing zone	Available husk from rice mill, '000' tonne/yr	Potential of electricity generation	
		Steam turbine plant kW	Gasification plant kW
		@1.3 kg husk/kWh$_e$ *	@1.86 kg husk/kWh$_e$ **
Dinajpur	82.90	7250	5150
Ishawrdi	121.21	10750	7500
Bogra	58.69	5200	3600
Naogaon	192.55	18250	12800
Total	**455.35**	**41450**	**29050**

*[8], **[7]

Husk supply from rice mills is not uniform in quantity throughout the year. Sometimes the husk supply is surplus so that the husk is leftover after consumption and sometimes the husk supply is deficit than needed. Therefore, further analysis is done to see the surplus and deficit pattern of husk supply throughout the year. In this study it is estimated that the husk supply is surplus for 8 months and the quantity of surplus amount is 19428 tonne for Dinajpur. This surplus amount of husk is to be stored for the use to overcome the deficit of rice husk for rest of four months of the year. The deficit amount is 17956 tonne, therefore surplus amount of husk is sufficient to overcome the deficit amount for 7250 kW steam turbine power plant at Dinajpurt area (Fig. 1). Similarly 18540 tonne of surplus husk is sufficient to make up 18399 tonne shortage of husk for gasification power plant at Dinajpur (Fig. 2).

At Ishawrdi area, the husk supply is surplus for 9 months and the quantity of surplus amount is 16271 tonne. This surplus amount of husk is to be stored for the use to overcome the deficit of rice husk for rest of three months of the year. The deficit amount is 15806 tonne, therefore surplus amount of husk is sufficient to overcome the deficit amount for 10750 kW steam turbine power plant (Fig.3). Similarly 16432 tonne of surplus husk is sufficient to make up 15752 tonne shortage of husk for 7500 kW gasification power plant (Fig. 4).

At Bogra area, the husk supply is surplus for 9 months and the quantity of surplus amount is 2759 tonne. The deficit amount is 2474 tonne, therefore surplus amount of husk is sufficient to overcome the deficit amount for 5200 kW steam turbine power plant (Fig. 5). Similarly 3174 tonne of surplus husk is sufficient to make up 2335 tonne shortage of husk for 3600 kW gasification power plant (Fig. 6).

At Naogaon area, the husk supply is surplus for 9 months and the quantity of surplus amount is 23235 tonne. Whereas the deficit amount is 21299 tonne, therefore surplus amount of husk is sufficient to overcome the deficit amount for 18250 kW steam turbine power plant (Fig. 7). Similarly 22697 tonne of surplus husk is sufficient to make up 21477 tonne shortage of husk for 12800 kW gasification power plant (Fig. 8).

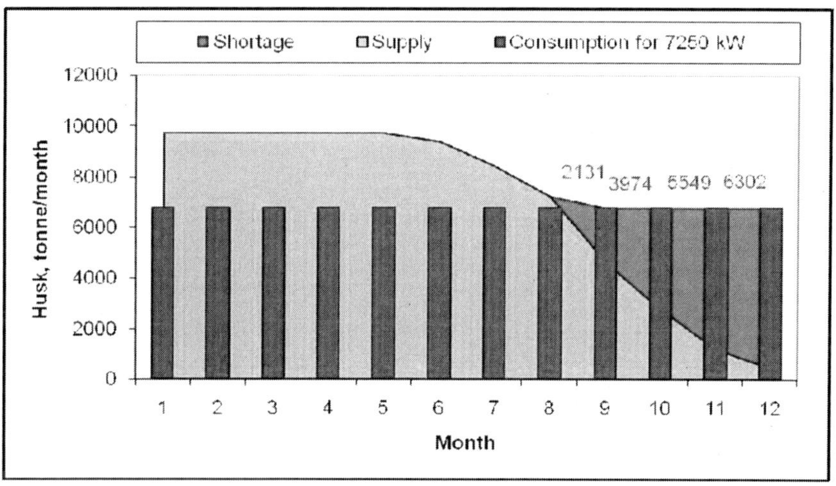

Figure 1. Potential production capacity of electricity at Pulhat for steam turbine plant

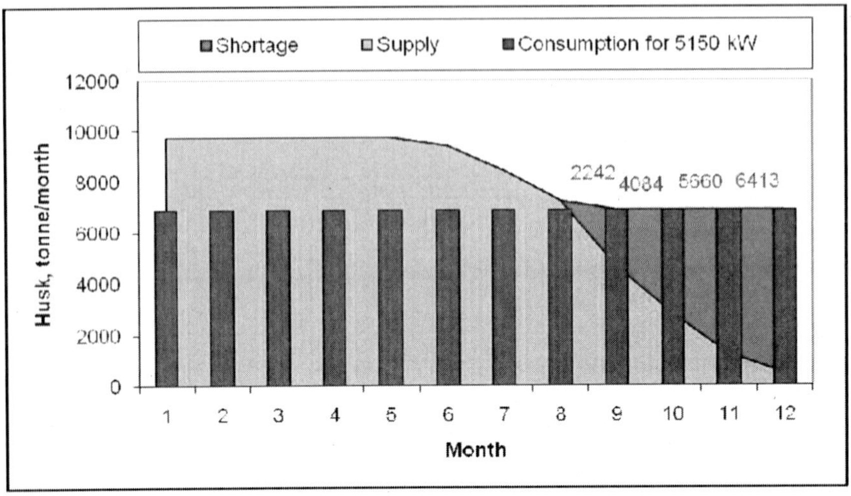

Figure 2. Potential production capacity of electricity at Pulhat for gasification plant

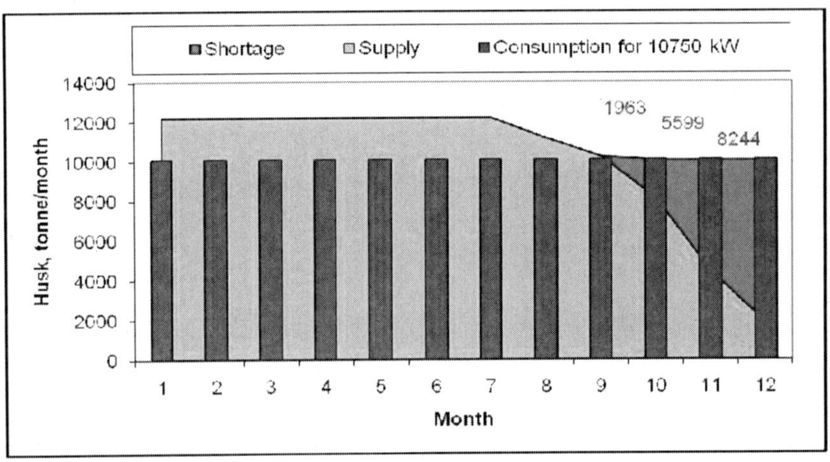

Figure 3. Potential production capacity of electricity at Ishawrdi for steam turbine plant

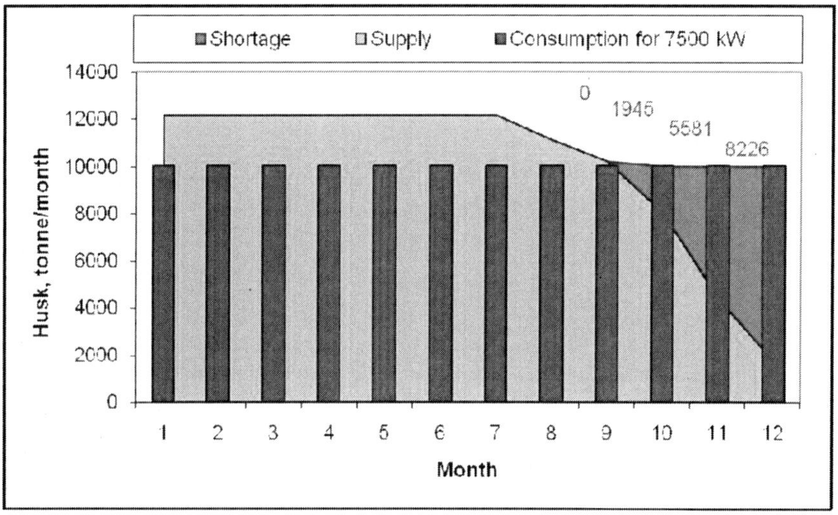

Figure 4. Potential production capacity of electricity at Ishawrdi for gasification plant

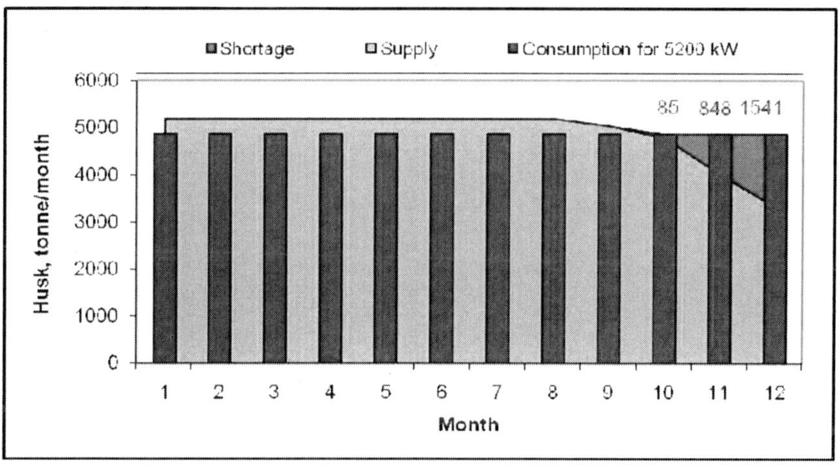

Figure 5. Potential production capacity of electricity at Bogra for steam turbine plant

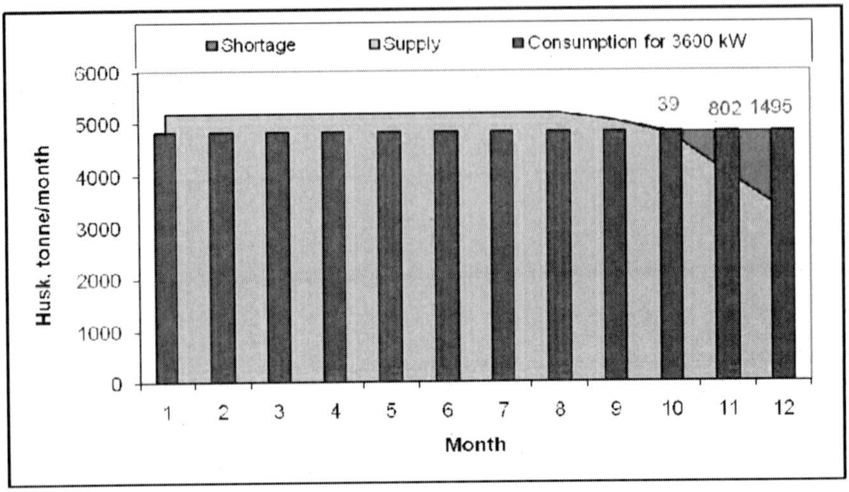

Figure 6. Potential production capacity of electricity at Bogra for gasification plant

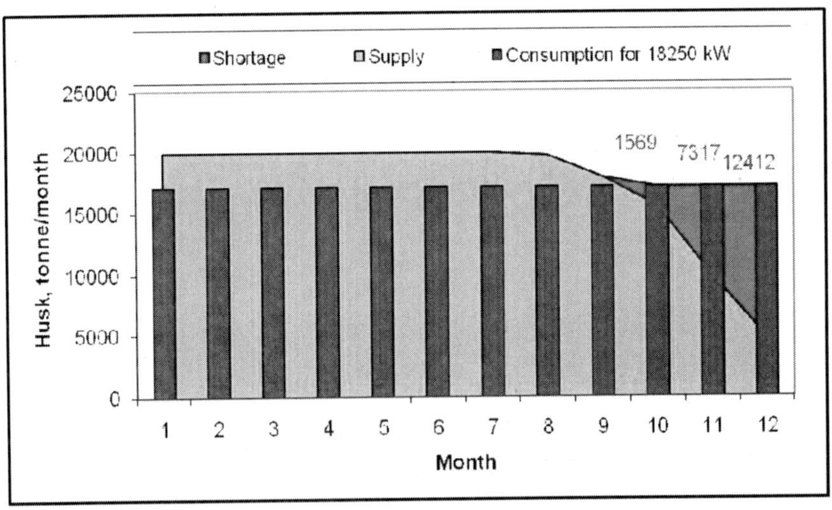

Figure 7. Potential production capacity of electricity at Naogaon for steam turbine plant

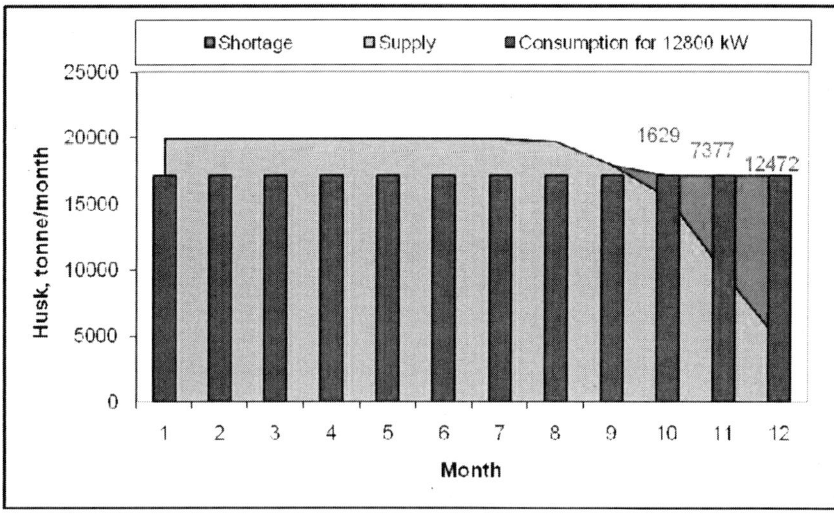

Figure 8. Potential production capacity of electricity at Naogaon for gasification plant

3.3. Projection Of Future Rice Production In Bangladesh

The production of rice is projected using the growth of 2.35%. The growth is calculated from the average growth of rice production data from 1971 to 2009. The projected rice production is increased to 71 million tonne in 2030 (Fig 9).

3.4. Rice Processing In Bangladesh

All produced paddy is not processed for preparation of food grain. About 5% of total paddy is used as seed for using next growing season. Among the rest of paddy about 70% is processed in local rice mill. The rest 25% of paddy is processed in rural household level. Rice husk is produced as by product in rice mill at the rate of 20% of mass fraction of paddy. The rice husk is primarily consumed for steam producing in parboiling process and drying process of paddy. The rest amount of husk could be an important source of husk based electricity generation plant.

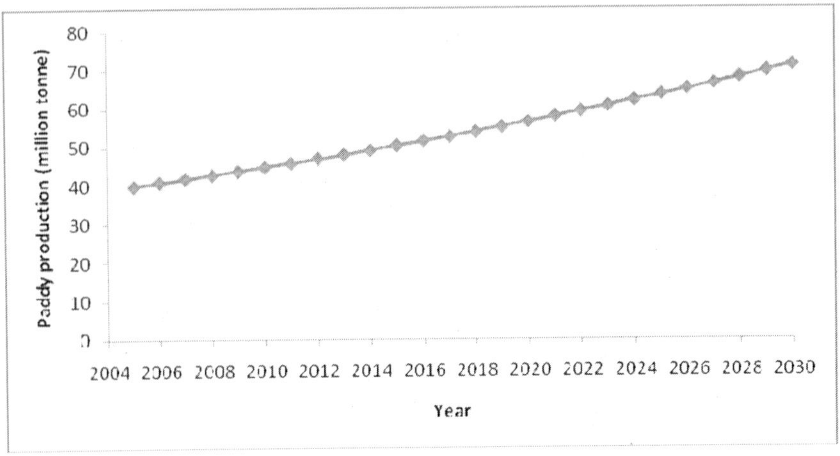

Figure 9. Raw rice production scenarion in Bangladesh

3.5. Husk Produced In Rice Mill

Amount of husk produced in rice mill depends on the amount of paddy processed in mill. Husk produced in rice mill could be increased by increasing the amount of processed paddy in mill. Share of paddy processed in mill is considered 90% of total paddy at the end of 2030 for scenario-1. As a result the husk production is increased from 9.96 million tonne (reference scenario) to 13.16 million tonne (Fig 10).

3.6. Husk Consumed In Rice Mill

Husk is consumed primary for steam producing in rice parboiling process. More amount of parboiled paddy consumes more rice husk. At present about 90% of paddy is parboiled process. Therefore major amount of husk is consumed for parboiling of paddy. About 120 kg of husk is consumed for each tonne of paddy parboiled. The husk consumption for parboiling could be reduced from 120 kg to 49 kg for each tonne of paddy parboiled by replacing the traditional boiler with improved boiler. At present all boilers used are traditional type. In scenario-1, it is assumed that 95% of traditional boiler is to be replaced by improved boiler to reduce the husk consumption for parboiled process. The other way to reduce the rice husk consumption is reduction of parboiled coverage as well as enhancement of unparboiled process. The results of husk consumption for two different scenarios are shown in Fig 11. The husk consumption is reduction to 4.56 million tonne (Scenario-1) from 6.47 million tonne (reference).

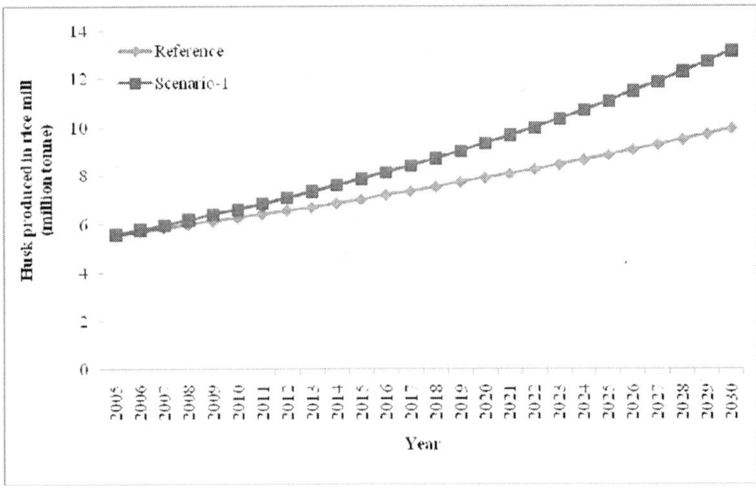

Figure 10. Rice husk production scenario in rice mill

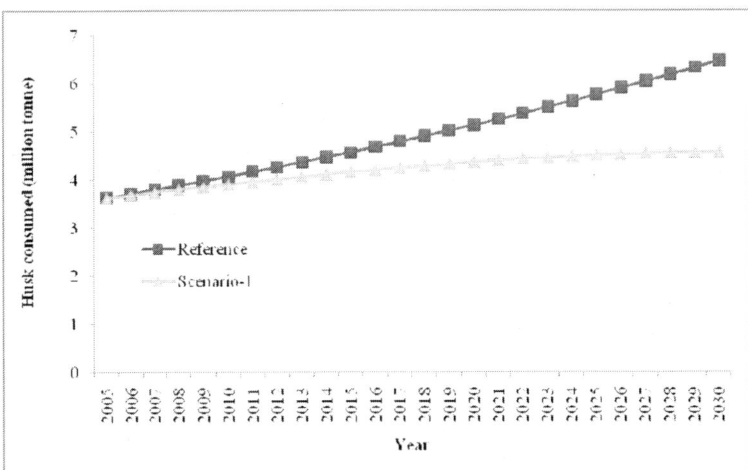

Figure 11. Rice husk consumption in rice mill at reference case ad scenario-1

3.7. Husk Surplus After Consumed In Rice Mill

The surplus amount of husk is calculated deducting the consumed amount of husk from total potential of husk (20% of paddy mass). The surplus amount of husk is increased to 8.60 million tonne (Scenario-1) from 3.49 million tonne (Reference) (Fig 12). The results show that there is huge potential to save rice husk by changing the technology for parboiling.

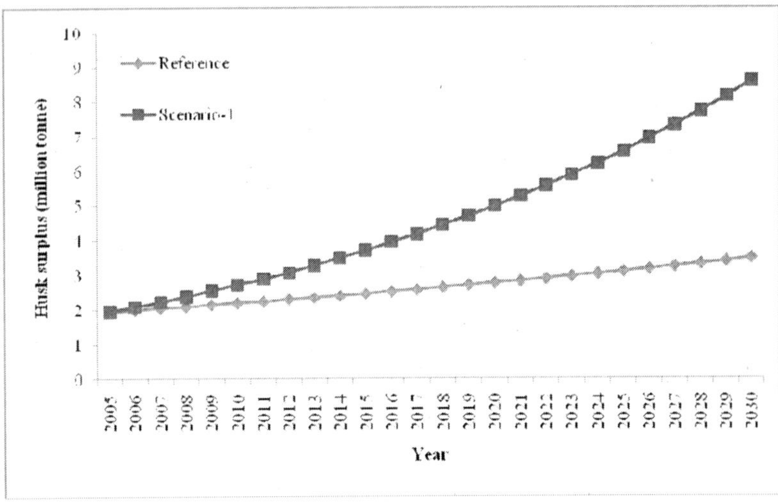

Figure 12. Rice husk surplus from reference case and scenarios-1

3.8. Potential Of Electricity Generation From Rice Husk

Rice husk biomass easily can be used for electricity generation. Amount of electricity generated from rice husk depends on the amount of raw material and type of technology used for energy conversion. There are several technologies for electricity generation from biomass. In the present study steam turbine and gasification technology are considered for estimating the potential of electricity generated from surplus amount of husk from different scenarios. For a steam turbine power plant consumption of rice husk is 1.3 kg per kWh electricity as reported by Singh [7]. On the other hand for a gasification power plant consumption of rice husk is 1.86 kg per kWh electricity generation [8]. Based on this assumption the potential electricity generated from husk is presented in Fig 13 and Fig 14.

Potential electricity generated is estimated to be 1689 GWh (Reference), 2092 GWh (Scenario-1) in 2010 and these values will be increased to 2683 GWh (Reference) and 6612 GWh (Scenario-1) in 2030 for steam turbine technology use (Fig 13).

Potential electricity generated is estimated to be 1180 GWh (Reference) and 1462 GWh (Scenario-1) in 2010 and these values will be increased to 1875 GWh (Reference), 4621 GWh (Scenario-1) in 2030 for gasification technology use (Fig 14). Corresponding power capacities of husk based generation plant are shown in Fig 15 and Fig 16.

Electricity consumption for rice processing in Bangladesh is shown in Fig 17. Amount electricity consumed is estimated to be 812 GWh (Reference) and 891 GWh (Scenario-1) in 2010 and these values will be increased to 1292 GWh (Reference) and 2030 GWh (Scenario-1) in 2030. The results show that amount of electricity generated from rice husk is sufficient to meet the electricity needed for rice processing purposes. Moreover, surplus amount of electricity could be served to national grid.

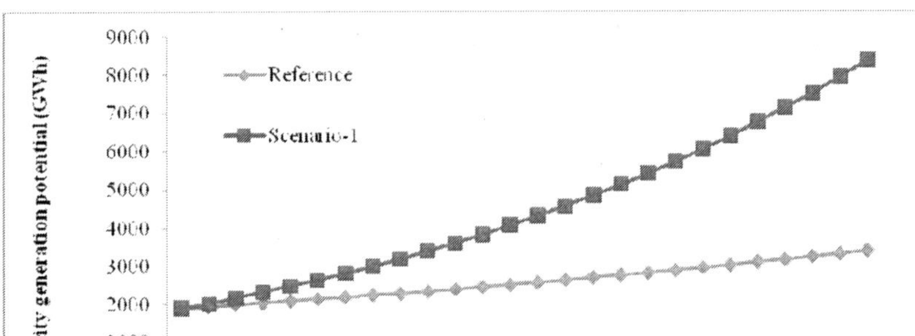

Figure 13. Potential of electricity generation from available rice husk at different scenario for steam turbine technology

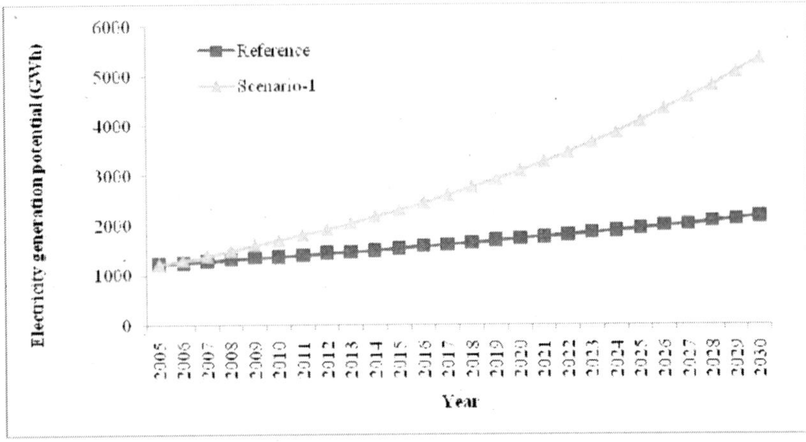

Figure 14. Potential of electricity generation from available husk at different scenarios for gasification based technology

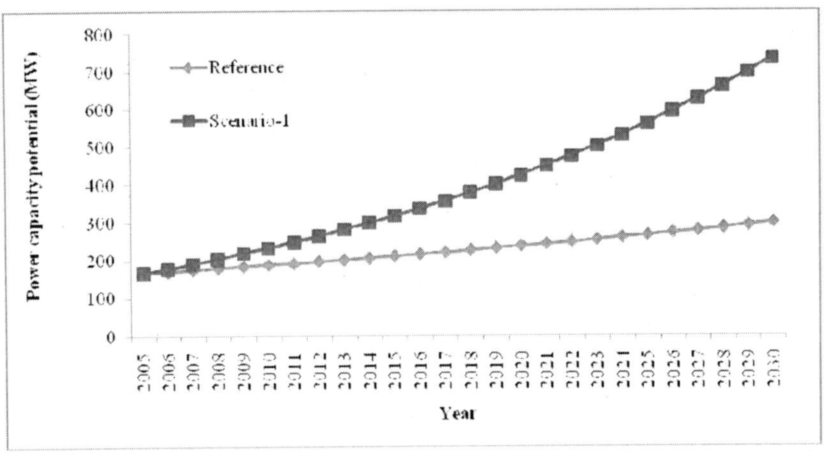

Figure 15. Potential of power capacity from available rice husk at different scenarios for gasification based technology

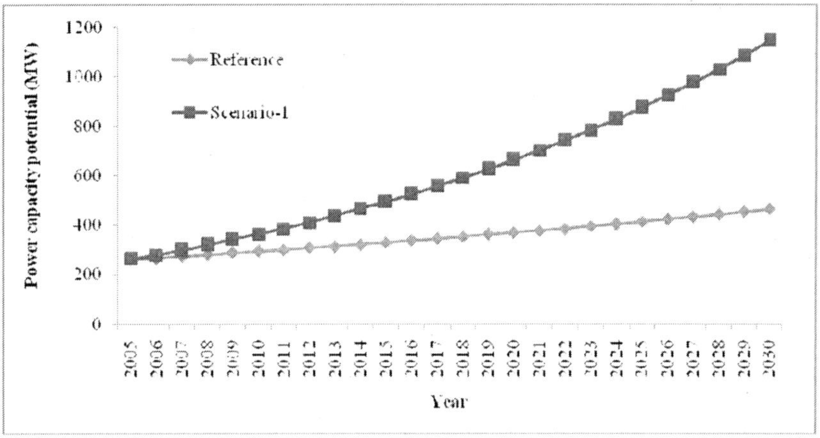

Figure 16. Potential of power capacity from available rice husk at different scenario for steam turbine technology

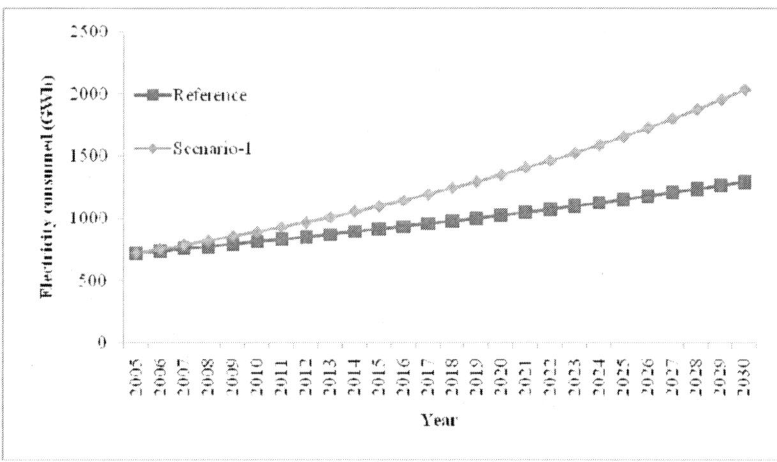

Figure 17. Electricity consumption for rice milling at different scenario

4. CONCLUSION

Power generation from rice husk is modern use of this waste material. The study results conclude that there is a potential to produce electricity from rice husk in selected rice processing zones of Bangladesh. It is estimated that 29 to 41 MW capacity generation plant can be installed in the selected areas based on the available surplus husk after their own consumption. In national context maximum 7682 GWh electricity can be generated from rice husk with a power capacity of 1066 MW in 2030. The amount of electricity generated from rice husk is sufficient to meet the electricity needed for rice processing purposes. Moreover, surplus amount of electricity could be served to national grid.

REFERENCES

1. FAOSTAT (2011) http://faostat.fao.org/site/339/default. aspx accessed on 20.8.11

2. BBS (2009). Statistical yearbook of Bangladesh. Dhaka: Bangladesh Bureau of Statistics, Ministry of Planning.

3. Ahiduzzaman, M (2007). "Rice Husk Energy Technologies in Bangladesh" Agricultural Engineering International: the CIGR Ejournal. Invited Overview No. 1. Vol. IX. January, 2007.

4. Ahiduzzaman, M., Baqui, M. A., Tariq, A. S. and Dasgupta, N. (2009). Utilization of Rice Husk Energy for Rice Parboiling Process in Bangladesh. Intl. J. BioRes 6(2):47-79.

5. Baqui, M. A., Ahiduzzaman, M., Khalequzzaman, M., Rahman, M. M., Ghani, J. and Islam, S.M.F. (2008): Development and Extension of Energy Efficient Rice parboiling Systems in Bangladesh. A Comprehensive Research Report submitted to German Technical Cooperation (GTZ).

6. Ahiduzzaman, M. and Islam, AKM S (2009). Energy Utilization and Environmental Aspects of Rice Processing Industries in Bangladesh. Energies 2009, 2, 134-149;

7. Singh, R. I. (2007). Combustion of Bio-Mass in an Atmospheric Fbc: An Experience & Study. Paper presented at the International Conference on Advances in Energy Research Indian Institute of Bombay, December 12-15, 2007

8. Islam, K. (2008), Senior Adviser, SED project, GIZ, Dhaka.

Chapter 6

An Overview of Financial Aspect for Thermal Power Plants

Soner Gokten[1]

[1] Gazi University, Turkey

1. INTRODUCTION

Thermal Power Plants are facilities that produced electrical energy which is a secondary energy source by using the primary energy sources. In reference to International Energy Outlook Report, marketable energy consumption in the World, taking 2007 as the base year, is expected to grow 49 percent until 2035 (IEO, 2010). According to this report, the distribution of estimated energy consumption by primary sources is shown in Fig. 1.

The revealed projection provides a picture concerning what kind of primary sources of energy should be appealed on producing electricity. It is put forward that the production of electricity with an increase of 87 percent until the year 2035 will respectively reach up to 25 trillion in 2020 and to 35.2 trillion kilowatt hours in 2035, taking 2007 as the base year in the report. As can be seen in Figure 2, coal is primary energy source, the most preferred, in production of electricity and natural gas, renewable energy sources and liquid energy sources, respectively, follow it. At present, 80 percent of the World's production of electricity is carried out by fossil fuel power plants (coal, petroleum products, natural gas), 20 percent of the

World use the different types of primary energy source like hydraulic, nuclear, wind, solar, geothermal and biogas (SRWE, 2007).

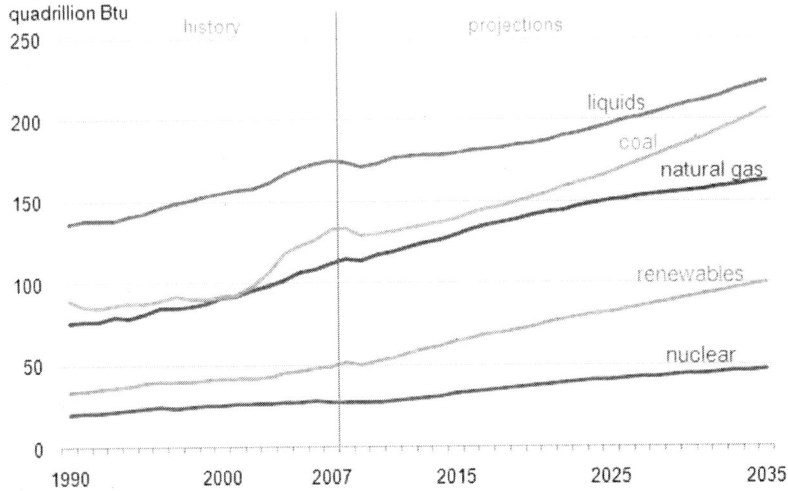

Figure 1. World marketed energy use by fuel type, (IEO, 2010).

Undoubtedly, the choice of the primary sources for electricity production depends on many factors and these factors are fundamental to determine the types of thermal power plants. Thus, these factors also drivers of the decision making processes of financial management function in thermal power plants. Also, the close relation, between financial development and economic growth, increases the importance of financial perspective in management of thermal power plants.

Financial management decision making process is applied on the basis of present value maximization within the framework of the investment and operating costs. Therefore, the factors, influencing decisions related to thermal power plants, affect investments and operating costs indirectly. In this study, it is tried to examine the components that related with financial decision and to present the internal relations between the financial management and thermal power plant in a panoramic view. In the following sections, financial decision making process in thermal power plants are discussed, under the titles of 'The Factors Affecting Decision

Making Processes', 'Investment, Operating and Supporting Costs' and 'Long-Term Supply Contracts' respectively. Concluding remarks are then offered.

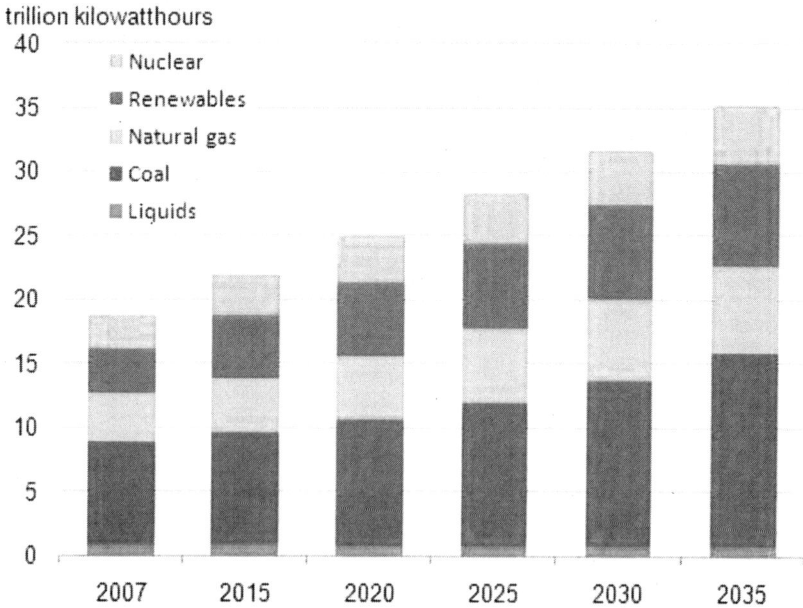

Figure 2. World net electricity generation by fuel, (IEO, 2010).

2. THE FACTORS AFFECTING DECISION-MAKING PROCESSES

Energy sector is a dynamic market that contains many guiding factors. Human beings, especially after industrialization, understood the vital importance of the sector and became to mobilize its ability to obtain energy. Playing such a critical role of energy sector made inseparable part of public sector for the developed and developing countries. Namely, energy sector must not be evaluated on the basis of market economics. This kind of effort will not reflect the truth. Because the invisible hand of public is always on the sector, while the private sector rules seem working.

Thermal power plants is one of the most important element of the energy sector and they are masterworks that enable producing electrical energy which can be thought as one of the basic needs of life after water and food. Preference of the thermal power plant's type in electricity production is a big dilemma and prior discussion subject for related parties in recent years. For instance, environmentalists act against fossil-fuelled thermal power plants or nuclear power plants and they try to warn decision-makers about environmental pollution, global warming, carbon emissions etc. There is no doubt that eliminating the existence of such kind of industrial elements, playing a major role in environmental issues seems true, but only with a view of environmentalism because when many other factors were considered, thoughts of environmentalists cannot be accepted in short- or medium-term. The thoughts, agreed and supported by everyone, can be ignored suddenly and quickly because the needs, called energy, especially in economic frame, have a vital importance. Financially; a bird in the hand is better than two in the bush.

All decisions on the type, innovation and improvement efforts of thermal power plants depend on many factors. It is impossible to execute decision making process as taking only one of these factors. The effective factors in the establishment and operation decision of thermal power plants can be classified as follows;

- Government Policies and Preferences
- Environmental Factors
- Macroeconomic Factors
- Research & Development Opportunities

2.1. Government Policies and Preferences

Countries compete with each other within the framework of the energy production nearly. Energy, especially for the developed countries, is one of the most important factors within the frame of competitive advantage. The energy consumed by the countries is illustrated in the figure below.

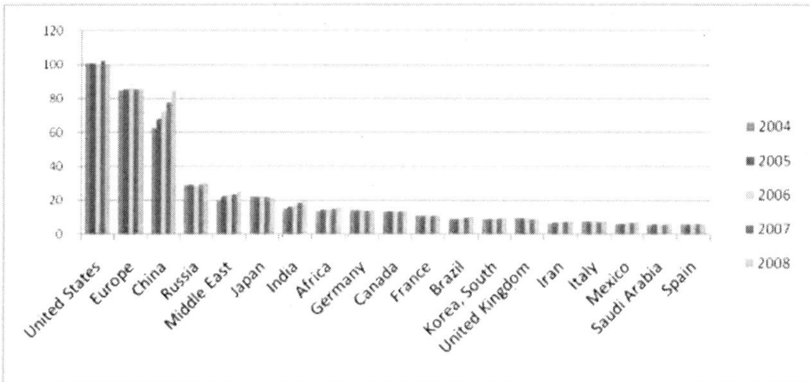

Figure 3. Countries by energy consumption (Quadrillion Btu), (EIA).

The close relations between industry and energy consumption, especially when looking the data of China, can be observed easily. Namely, energy consumption of Republic of China increased every year from 2004 to 2008. China industry, increasing production capacity, is reportedly building an average of about one coal-fired thermal power plant a week to meet the requirement of its high-level of electrical energy consumption (Figure 4) (Bradsher and Barboza, 2006).

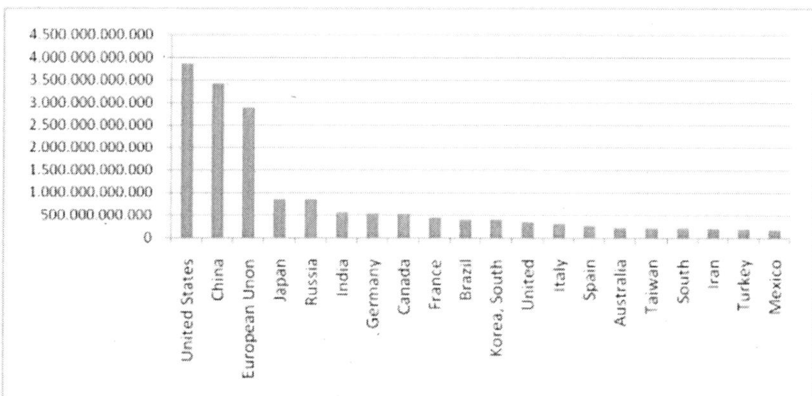

Figure 4. Countries by electric energy consumption (kwh), (EIA).

Primary energy source possibilities of countries are one of the basic factors that determine the preferences of a thermal power plants. Namely, U.S.A., Germany, India and China produce more than 50% of their electrical energy by coal-fired thermal power plants, while most of the thermal power plants, in the countries has an abundance of natural gas such as Qatar, are gas-fired. This choices are directly related with the reserve capabilities of the primary energy sources which are one of the main issues for government policies and preferences. For example, the coal-fired thermal power plants are preferred to operate reserves, without taking into consideration of performance criteria and environmental factors, although coals has a poor heat value and high carbon content in Turkey.

2.2. Environmental Factors

The basic element with energy production and environment interaction is the danger of climate change. Carbon dioxide, methane, nitrous oxide etc. is piling up on the layers of atmosphere, continuing of this situation by increasing prevents the reflection back and faced to a heating risk of the earth. Melting of glaciers and rising sea water are the natural results of warming phenomenon.

Undoubtedly, to tackle with the kind of this problem requires a global consciousness and joint actions. In this direction, the studies begin at the end of the 1980s, were brought to the agenda of the world in 1992 with Rio summit and a joint consciousness was formed by courtesy of the United Nations Framework Convention on Climate Change, accepted on 21st March, 1994. The UNFCCC, embraced a good faith at the point of decreasing its greenhouse gas oscillation based on 1990 by the parties in the contract. However, it seems insufficient because of the poor sanctions and absence of quantitative targets. Kyoto Protocol, accepted in 1997, 3rd Parties Conference of The Agreement of United Nations Framework Convention on Climate Change and came into operation in 16th February, 2005 is the first concrete step to achieve the ultimate purpose of the contract. Since the protocol includes concrete quantitative targets for the period of 2008-2012 and open sanctions, it creates possibilities for other international agreements to conflict with climate change.

The major role of energy sector in greenhouse gas oscillation as a principal cause of climate change; the measures, to be taken and apply against to such danger, is necessitated predominantly in energy sector. At this point, the most serious press is dragooned by environmental protectionist about fossil fuelled thermal power plants. According to them, fossil fuels should use just a bit in indispensable situation or new substances must be exist for taking substitute. There is no doubt that coal-fired thermal power has the highest rate of carbon emissions among the fossil fuelled thermal power plants. Many countries which want to use their primary energy sources in a most efficient way; invest a lot of money for technological research to develop clean coal, line to keeping carbon emissions, arrested, confinement, isolation, treatment, swallowed, destruction and storage. Several alternative methods investigated continuously (Prisyazhniuk, 2008). Many important leaders reckon on clean coals technologies in order to continue coal-fired based electricity production.

2.3. Macroeconomic Factors

The close relationship between economic development level of countries and energy or electric energy consumption is one of important elements which directly effects necessity of thermal power plant. There are a lot of works in literature that try to analyze the relationship between energy or electric energy consumption and economic growth.

The relationship between economic growth and energy consumption was found as casual for Japan byErol and Yu (1987), for Turkey, France, Germany, Italy, and Korea by Soytas and Sari (2003), for Canada by Ghali and El-sakka (2004), for Taiwan by Holtedahl and Joutz (2004). Lee and Chang (2008) found unidirectional causal relationship in the long-term between energy consumption and economic growth for 16 Asian Countries between 1971-2002 in their works. Moreover, in studies ofLee et al (2008), it was worked on 22 OECD countries base on 1960-2001 period, and as a result, a bi-directional causal relationship was found between energy consumption, capital, and GDP. Confessed works are very important, especially, for policy makers. Because, way of relationship between energy consumption and economic growth represents various options to decision-makers related to interaction between energy and GDP. Important works related to determination of the relationship between energy consumption and economic growths were represented chronologically in Table 1. These works, at the same time, suggest to be understood what energy dependency is in economically.

Mentioning about energy dependency needs to accept energy consumption as the most important part of economic growth. In other words, we add energy consumption between capital and work elements which are basic inputs of economic growth and accept that a decline in energy consumption effects real GDP negatively. To confirm this kind of hypothesis, causality must be realized as runs from energy consumption to economic growth. On such an occasion, country economy is expressed dependent on energy, and decision-makers try to prevent the negative effect on real GDP to apply energy conservation policies.

On the contrary, if the unidirectional causality runs from real GDP to energy consumption, decrease in the energy consumption will not affect real GDP significantly. If bi-directional causality happens between energy consumption and economic growth, energy consumption and real GDP effects each other simultaneously. This kind of interaction directs policy makers to take measures on energy use. But, at the point of declining energy use in arrangements, it is necessary to separate industrial dimension of energy use to avoid from potential negative effects of energy consumption on real GDP. In this frame, decision-makers can assume that to head for more efficient energy sources or to use energy generating technologies which creates less environmental pollution as an alternative.

The basic relationships, mentioned above, represent a macroeconomic point of view related with which the reasons the countries faced to energy dependency or had bi-directional causality between economic growth and energy consumption, prefer thermal power plant. Because, each country is not able to fulfil the Kyoto Protocol completely and it seems impossible to quit thermal power plants with fossil fuel for some countries in economic frame. Primary sources are the most important inputs for production process in thermal stations. For this reason, fuel costs are important economic factors for coal, nuclear, and natural gas based thermal power plants. The thermal power plants, use solar, geothermal, and wind energy, are excluded because of none of fuel cost. Historical costs related to coal and natural gas costs are represented in the figure 5 and 6 respectively. Uncertainties of costs of primary sources using as a fuel in electric production have critical effects on policies that are applied by the countries in energy sector. The easiest way to manage mentioned uncertainties is to build thermal power plants according to domestically existing primary sources of the countries. Otherwise, management effectiveness cannot be provided in the relationship between energy consumption and economic growth.

Table 1. Overview of Selected Studies, (**Belke et al., 2010**).

Researcher(s)	Method	Countries	Result
Kraft and Kraft (1978)	Bivar. Sims Causality	USA	Growth — Energy
Yu and Choi (1985)	Bivar. Granger test	South Korea	Growth — Energy
		Philippines	Energy — Growth
Erol and Yu (1987)	Bivar. Granger test	USA	Energy ↛ Growth
Yu and Jin (1992)	Bivar. Granger test	USA	Energy ↛ Growth
Masih and Masih (1996)	Trivar. VECM	Malaysia, Singapore & Philippines	Energy ↛ Growth
		India	Energy — Growth
		Indonesia	Growth — Energy
		Pakistan	Energy — Growth
Glasure and Lee (1998)	Bivar. VECM	South Korea & Singapore	Energy — Growth
Masih and Masih (1998)	Trivar. VECM	Sri Lanka & Thailand	Energy — Growth
Asafu-Adjaye (2000)	Trivar. VECM	India & Indonesia	Energy — Growth
		Thailand&Philippines	Energy — Growth
Hondroyiannis et al. (2002)	Trivar. VECM	Greece	Energy — Growth
Soytas and Sari (2003)	Bivar. VECM	Argentina	Energy — Growth
		South Korea	Growth — Energy
		Turkey	Energy — Growth
		Indonesia & Poland	Energy — Growth
		Canada, USA & UK	Energy — Growth
Fatai et al. (2004)	Bivar. Toda and Yamamoto (1995)	Indonesia & India	Energy — Growth
		Thailand&Philippines	Energy — Growth
Oh and Lee (2004b)	Trivar. VECM	South Korea	Energy — Growth
Wolde-Rufael (2004)	Bivar. Toda and Yamamoto (1995)	Shanghai	Energy — Growth
Lee (2005)	Trivar. Panel VECM	18 developing nations	Energy — Growth
Al-Iriani (2006)	Bivar. Panel VECM	Gulf Cooperation C.	Growth — Energy
Lee and Chang (2008)	Mulitv. Panel VECM	16 Asian countries	Energy — Growth
Lee et al. (2008)	Trivar. Panel VECM	22 OECD countries	Energy — Growth
Narayan and Smyth (2008)	Multiv. Panel VECM	G7 countries	Energy — Growth
Apergis and Payne (2009)	Multiv. Panel VECM	11 countries of the Commonwealth of Independent States	Energy — Growth
Apergis and Payne (2009)	Multiv. Panel VECM	6 Central American countries	Energy — Growth
Lee and Lee (2010)	Multiv. Panel VECM	25 OECD countries	Energy — Growth

Countries which have high-level coal reserve still prefer coal-fired thermal power plants in primary level despite with environmental problems. Costs of coals, used as fuel are directly dependent on calorie rate which determines heat capacity of coal. Nonetheless, the inorganic elements within coal caused environmental pollution. Anthracite has highest heat rate and least ash and poisonous exodus for environmental pollution, is a coal. Nonetheless, reserve size of it is much less than the reserve size of lignite because, it's history bases on 300 million years ago. Water and a lot of foreign substance exist within lignite, approximately started to form 60 million years ago. Therefore, it shows a increasing feature of environment pollutions.

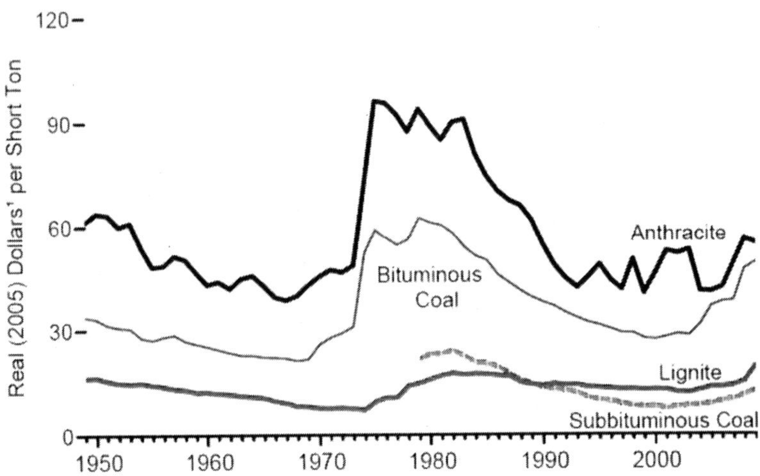

Figure 5. Coal prices by type for the period of 1949-2009, (AER, 2009).

As it has been seen in the figures, natural gas price which is used as fuel in electric production, is especially more unstable than coal price. Especially, lignite price has almost been stable during years. It will be able to be understood why coal is more preferable in energy production, if it is taken into consideration that coal price is much less than natural gas cost. Leaving coal-fired thermal power plants, because of environmental pollution, for a country which has high-level lignite reserve, is as unreasonable as depending on an uncontrolled price formation while built gas-fired thermal power plant for a country which has not enough natural gas reserve. Dilemma in this example becomes more complex when all factors are handled together about thermal power plant management and its construction.

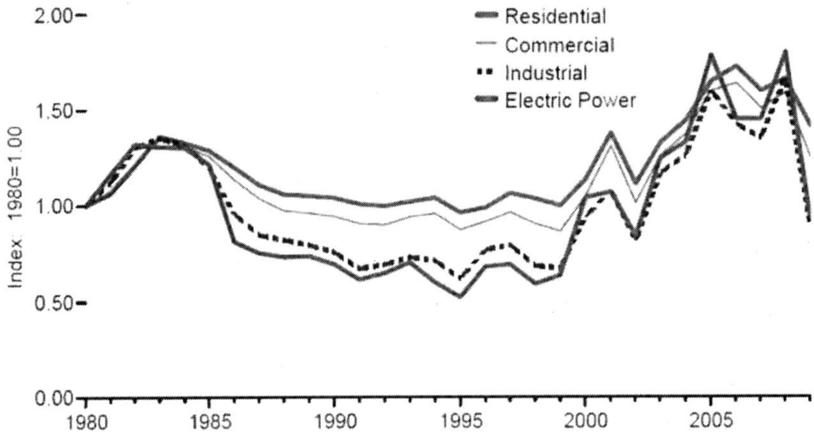

Figure 6. Natural gas indexed prices by sector for the period of 1967-2009, (AER, 2009).

2.4. Research & Development Opportunities

The Thermal power plants, besides being the facilities producing electricity, are one of the towering sectors in the world with the technologies they have. Countries positioning their research and development strategies in the thermal power plant construction and formation frame work in a right way derives revenue by selling the know how knowledge they got to other countries. In this framework, countries sometimes pioneer various thermal power plant investments for creating research and development opportunities and bringing the related technologies in their countries or they prompt the related field by making required arrangements related with decisions of thermal power plant type in their countries.

Nuclear Power plant is in the position of an important laboratory especially for the countries wanting to own nuclear technology. Up until now, many countries utilize the nuclear power plants as a device for the studies of developing nuclear technology or they swell the number of them.

Countries that can observe the future in a right way, namely the ones having vision, seriously make investments to research and development activities on types of thermal power plants, moreover they try to switch the energy production in their countries to the stated technology in spite of the potential damage risks. One of the most important sample of this is Spain.

In Spain, among the towering leaders in renewable energy field, important studies are made especially in solar energy field. Forasmuch as, Spain is in the position of an important producer and exporter country in photovoltaic industry.

At the root of this success of Spain, there is its displaying the special importance it gave to renewable energy sources with Renewable Energy Plan considering the period between 2005-2010 as a government policy and paving the way of research and development activities by switching solar thermal power plants. In accordance with Renewable Energy Plan big importance is given to developing photovoltaic batteries and also producing energy with this way. According to this plan, supporting R&D activities that develop the evolution, producing, trading and setting periods of photovoltaic battery technologies is decided. Spanish government gives the purchase guaranty of the electricity produced from the renewable energy sources and subsidizes in system installations.

Another field in which research and development studies gain speed is also wind power based thermal power plants. Germany, a world leader in wind energy, belongs to approximately one –third of total production capacity of the world. Australia, Canada, China, France, India, Italy, Philippines, Poland, Turkey, England, and USA can be showed as the other countries that can lead utilizing the wind power in the world.

3. INVESTMENT, OPERATING AND SUPPORTING COSTS

We tried to examine the factors affecting the decisions on thermal power plants: 'Government Policies and Preferences, Environmental Factors, Macroeconomic factors and Research and Development Opportunities'. Appearance of stated effects on financial side realize via costs.

These costs can be handled into three groups;

- Investment Costs
- Operating Costs
- Supporting Costs

Decision process on need of thermal power plants, type selection, establishment decision, managerial factors, etc. are carried on in financial framework over the three cost group stated above. Each cost factors are affected from the factors, details of which we expressed, and jointly make out a financial bill to decision makers. Figure 7 illustrates that in what kind of interaction the financial point of view in decision process is happened.

3.1. Investment Costs

Investment cost covers engineering, procurement, construction, transmission and capitalized financing costs within. Engineering, procurement and construction costs include preparing the project of the stated facility, providing required equipment and materials and all kind of costs related with construction activities following the decision of establishing thermal power plant. Type of thermal power plant is the basic phenomenon determining the capacity of cost item occurring from these three factors. In general, Nuclear power plant is the type of thermal power having the highest investment cost and respectively coal-fired, wind or solar and at last row gas-fired thermal power plants follow this. Moreover, increases in price of construction materials as iron and steel that show raw material and semi finished material features, increasing demands on machines and equipments used in electricity production facilities, increases in occupational wages and lack of educated manpower recently cause continuously increasing in these three costs.

Transmission costs are obligatory investments of thermal power plant establisher in primary source supply and distribution of produced electricity framework. How to make the transmission especially in primary source supply framework must be meticulously appraised in feasibility studies. Namely, basic reason that the coal-fired thermal power plants are established close to coal mines, nuclear power plants are established in seaside and lakesides by reason of water need etc. is decreasing required transmission investment costs.

Financing costs endured for supply of required fund for establishing thermal power plant are included in investment cost by being capitalized in framework of feasibility studies. Interest and equity costs take place in it.

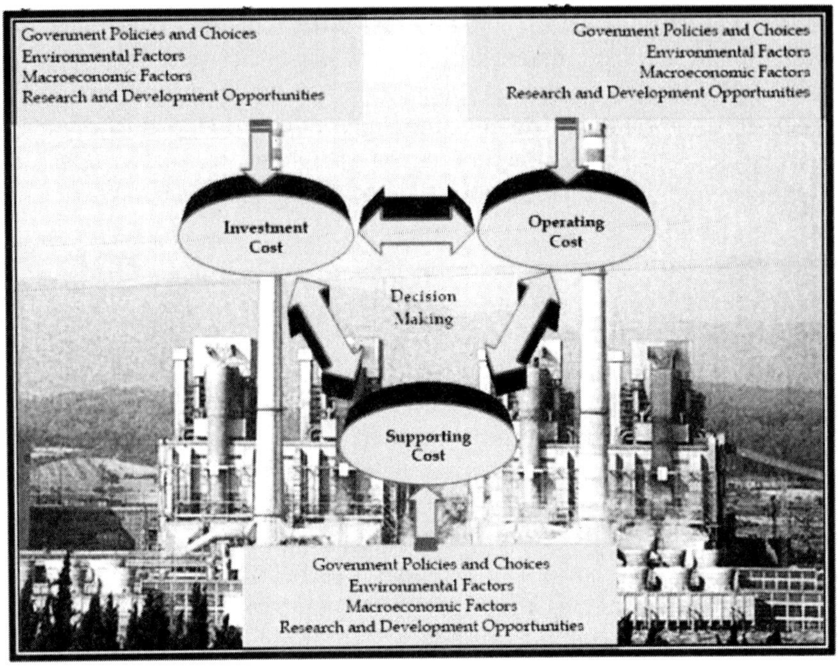

Figure 7. Interaction diagram of decision making process.

3.2. Operating Costs

Operating costs consist of expenditures that thermal power plant endures in order to fulfill the main activity. Stated costs are grouped in themselves as direct labor cost, direct material cost, manufacturing overhead cost, marketing cost, and administrative cost. Direct labour cost includes the wages directly paid to working staff, direct material cost includes the expenditures for raw and semi-finished materials, manufacturing overhead cost includes direct or indirect expenses on production. These three cost groups make the producer reach the unit production cost. In addition to this, as it can be understood from their names, marketing cost and administrative cost include the expenses on marketing and administrative functions. All costs can show fixed, variable or semi-variable features.

The most important cost item in operating costs of thermal power plants framework is the primary source of energy, in other words fuel expenses. Fuel expenses are in direct material costs. While the ratio of stated cost for thermal power plants using natural gas and fuel oil in the total operating cost is around ninety percent, this ratio is remained around at the level sixty per cent to seventies for coal-fired thermal power plants. There is no fuel cost in the thermal power plants using renewable energy such as hydraulic, solar and wind energy.

Fuel cost means charging the production cost of the primary energy source cumulatively into the production cost of thermal power plants. Namely, for example, the coal price of coal-fired thermal power plants using coal mine also includes expenses endured for removal of the coal from the mine. For this reason, thermal power plants highly depend on reserve status, manufacturing activity, price formations of the primary energy source.

3.3. Supporting Costs

Among the factors affecting decisions on thermal power plants, costs formed depending on environmental, strategic, socio-political and socio-economic all kind of factors, not for the influence of affecting investment and operating costs, must be evaluated in this group.

Two above-mentioned groups are often introduced in thermal power plant decisions and supporting costs are ignored. Many cost-create events and elements as activities of environmentalists, revision of energy policy in view of natural disasters, building new facilities in order to induce to use clean coal because of environmental factors, supporting the R&D activities needed for developing an energy production technology that will be standing out in the future can be evaluated in supporting cost group. Two basic features of supporting costs can be mentioned: (1) These costs are generally invisible or less visible. In other words, it is difficult to take into an account, because it shows facility that can be calculated depending on several possible scenarios. Therefore, the effectiveness of the energy sector could be increased if they could be planned before they occur. For example; when it is started to establish coal-fired thermal power plants, in what extend is it thought to meet with Kyoto Protocol? (2) These costs can show a negative or a positive feature for the countries. In other words, it can turn into sunk cost according to the way of occurring the event and it can create yields. Spain's research & development costs spent for departure of solar thermal power plants, has not been sunk costs and started to turn into yields within the framework of exports. For this reason, all possible event that can occur while decisions of thermal power plants are being financially evaluated should be analyzed.

4. LONG-TERM SUPPLY CONTRACTS

The construction decisions of thermal power are taken in the frame of strategic plans and long-term supply contracts made between primary energy source producers and thermal power plant operators. The main reason for this is that each one of them wants to minimize risks depending on fluctuations in the spot market (Joskow,1987,1988,1990; Williamson, 2000).

If the members of long-term supply contracts are both state-owned companies, occurred disagreements between these companies can be settled according to directives of the authority. In this sense the contract price cannot reflect world prices. However, potential private investors seek a contract with powerful sanctions and governance mechanism for the exchange (Uner et al., 2008).

Here, we will try to search the answer of the question ' What kind of a long-term supply contract should be made between thermal power plant and the primary energy source, in condition of the fuel supplier?' Each of following sub-sections briefly describes the parts are to be included in long-term supply contracts.

4.1. Maturity Of Contract And Suspensory Conditions

Maturity of supply contracts can be formed as 5 years, 10 years and multiplies of them. However, because the supply contracts cover long periods, many substances taken place in the contract show revisable qualifications. The basic element is that the number of factors which may be needed to be revised will also increase in parallel with prolongation of the maturity of the contract. Because the longer-term increases uncertainties about the future, which increases the number of condition and requirement in making the value chain reached at an optimum level the framework of increasing the number of terms and conditions.

Depending on prolongation of supply contract, transforming many factors into revisable form is not an desired condition. The main reason for this is the paradox of making political risks increased, despite the minimization of operational risks increased. Also, increase in revisable factors will increase the numbers of correction and/or adaptation transactions. In the long term, each correction and adaptation work can have possible negotiation request its train. Negotiations are source of political risks and they prevent the job from being done in a systematically continuous way. For this reason, determination of maturity for long-term supply contracts takes place at the beginning of the most important decision items. It is desired in maturity decision that contracts which do not create political risks in parties, do not have any revisable factors more than required and take into consideration the possible stability period on market in which the transactions will be realized.

"First delivery date" taking place in the parts on long term supply contracts is accepted as maturity beginning of the contract. So, signing date expresses the starting point of the "first delivery date" in association of activities with maturity framework while giving effect to the contract in legal framework. Occations in which induring of contracts clauses are subjected are described as suspensory conditions. Betrayed obligation belonged to parties within the context of contracts should be fulfilled in the whole suspensory conditions as of the singing date. In other words, parties, must provide the conditions presented as suspensory conditions,during the contract. Suspensory conditions determined for seller exposes the qualification criteria of the seller for works and products.

Suspensory conditions determined for purchaser are formed especially according to the using aim of primary energy source mentioned in the contract. Such that, at the same time, for what the fuel is bought by purchaser, in which test and how it will be used and what kind of production process will it be met form the criteria that the purchaser must have. The importance of the suspensory conditions that will exposed for purchaser is that supply contract is signable before start of the work. Thus, that the work for which fuel will be used as raw material have high operating leverage makes necessary to produce solution for supply trouble before facility construction of the purchaser.Hence, purchaser is responsible for the facility that he will establish because he is to determine "the first delivery date " following the signing of the contract.

4.2. Sale And Purchase Obligations

Determination of moral rules of commertial processes in long term supply contracts is necessary to make supply relationship effective in long term. While responsibilities evaluated in Sale and Purchase Obligations concepts framework regulate alternative supply resources, using fields of the subject product of commertial processing and similar subject for purchaser; for seller, they regulate the points as stock control, supply mechanism.

Minimization of the operational risk possessed by continual production activity presents importance in long term supply contracts. According to this, enterprise which will use its fuel as a raw material will need continuousness of input to provide continuity in production. For this reason, the purchaser's amount of input stock must be follawable for seller, deliveries and supply contact network must be processed according to purchaser's stock. Long term supply contracts must guarantee the fuel need of purchaser. Especially, if fuel supply cannot be realized due to the negativenesses occurring except force majeure, providing the short term need of purchaser is needed. Stated necessity makes defraying the loss and similar arrangements obligatory.

4.3. Amount and Delivery Of Fuel

Amount of related fuel is called contract amount in long term supply contracts. Naturally, contract amount forms the important part of supply contracts.

Contract amount expresses total amount of fuel supplied during a whole working year. Calendar of working year can be determined either standard 12 months or less than 12 months. Hence, the amount of contract amount per a month is accepted as " planned amount-PM". So, contract amount is the main factor for determination of delivered fuel amount. Such that, contract amount is stable factor especially in planning the monthly deserving. In this context, in order to protect the parties' obligations, contract amount and planned amount do not change. In other words, amount of fuel supplied during the working year cannot be less than contract amount.

4.4. Fuel Quality

Clearly expressing the quality of fuel introduced in long term supply contract as well as maturity and pricing one of the most important factors of the contract.

Defining the quality factor as gaps instead of a clear definition make the effective practice of the contract possible. For example, three gaps can be determined for coal according to calorie value. From higher to lower, these can be described as (1) Incitement, (2)Normal, (3)Discount. Price is not revised in normal gap. It is increased in incitement gap (Premium); it is decreased in discount gap (fine). Thanks to this, an mutually acceptable price fixing for purchaser and seller is created.

4.5. Fuel Price

The most importing issue in the long-term supply contracts is pricing. Pricing can be done by using three approach, called: cost approach, market approach and income approach.

The necessity of certainly reflecting the production costs of the enterprise that produce the primary energy source to the price is the thing wanted to express by cost approach. It is reached to this price as a result of profit margin that will be included at the base price by using of income approach. Base price shows a price feature that can be fixed according to conjectural movements in the market and in this framework, it forms the beginning period price of the contract. Besides, it is corrected in the periods determined in the contract and correction is realized in market approach framework. At the base of the market approach, prices of all fuels that can be equal to the stated fuel and all other macroeconomic factors can be evaluated in correction of base price.

4.6. Payments and Collections

It is the part of long term supply contracts that organizes the decisions practiced by both parties as purchaser and seller. Delivering of the fuel in delivery point and doing the quality measurements make data flow obligatory together. So, preparing the bills show that the parties fulfil the obligations mutually in the consequence of the processes is highly important to provide this data flow.

Generally, Purchaser is obliged to advance money to the seller in exchange for specific ratio of the fuel and /or show bank guaranty. By way of requital, seller is obliged to deliver the fuel promised to the purchaser and document the process. After fulfilling the mutual obligations and finishing the contract, seller are to give the advance taken from purchaser and bank guaranty back.

4.7. Force Majeure

The most important feature of long term contracts is that each clause is adjudicated in usual, considerable conditions. Howsoever, realization of the activities and contracts planned in long term has risks within, initially political risk, as is due the structure of long term contracts, these risks are wanted to be lowered to minimal level.

Force majeure concept is one of the main concepts of law and practice of it is seen in all branches of law. Force majeure is an extraordinary, incidental event, fact, condition that blocks the fulfilment of a responsibility partly or entirely, permanently or temporarily, due to this feature cancelling or delaying the responsibility its fulfilment and maturity or changing the feature of the responsibility, unexpected and unpredictable; even if expected or predicted, that cannot be blocked. Even if in long term supply contracts precautions are taken for the risks that are possible to come across depending on ordinary conditions, it is possible to come across with extraordinary conditions.

5. CONCLUSION

Thermal power plants are the masterpieces realizing the production of electricity, sine qua non need of our world. The common aim embraced by everyone is preferring thermal power plants realizing the clearest production. However, various factors sometimes militate and they conduct the decision making process on thermal power plants contrary to the environmental sensations. It is needed to look the picture with a financial point of view to understand the reasons of this.

Nearly all of the decisions on type and administration of thermal power plants are made on gathering at the cost base in other words; at finance base. Therefore, it is needed to manage the factors affecting decision-making process for clearer electricity production, transform the supportive costs into investments that can create yields.

ACKNOWLEDGEMENTS

I would like to express my gratitude to Aslihan Akin and Eray Karaarslan who are from University of Turkish Aeronautical Association for their assistance.

REFERENCES

1. AER, 2009 Annual Energy Review 2009. Energy Information Administration, June 2009.

2. A. H. Belke, C. Dreger, H. Frauke, 2010 Energy Consumption and Economic Growth- New Insights into the Cointegration Relationship. Ruhr Economic Paper No. 190; DIW Berlin Discussion Paper No. 1017, June 1 2010.

3. K. Bradsher, D. Barboza, 2006 Pollution From Chinese Coal Casts a Global Shadow. The New York Times, June 11, 2006.

4. U. Erol, E. S. H. Yu, 1987 On the causal relationship between energy and income for industrializing countries. The Journal of Energy and Development, 1987, 13:113-123.

5. K. H. Ghali, M. I. T. El-sakka, 2004 Energy use and output growth in Canada: A multivariate co-integration analysis. Energy Economics, 26 2004 225 238 .

6. P. Holtedahl, F. L. Joutz, 2004 Residential electricity demand in Taiwan. Energy Economics, 26 2004 201 224 .

7. IEO, 2010 International Energy Outlook 2010. Energy Information Administration, July 2010, DOE/EIA-0484(2010).

8. P. L. Joskow, 1987 Contract duration and relationship specific investments. American Economic Review, 77 (1), 168-185.

9. P. L. Joskow, 1988 Price adjustment in long term contracts: the case of coal. Journal of Law and Economics, 31(1), 47 83.

10. P. L. Joskow, 1990 Price adjustment in long term contracts: further evidence from coal markets. RAND Journal of Economics, 21 (2),251-274.

11. C. Lee, C. Chang, P. Chen, 2008 Energy-income causality in OECD countries revisited: The key role of capital stock. Energy Economics 30(5), 2359 2373.

12. C. C. Lee, C. P. Chang, 2008 Energy consumption and economic growth in Asian economies: A more comprehensive analysis using panel data. Resource and Energy Economics, 30 50 65.

13. V. A. Prisyazhniuk, 2008 Alternative trends in development of thermal power plants. Applied Thermal Engineering 28 2008 190 194 .

14. U. Soytas, R. Sari, 2003 Energy consumption and GDP: causality relationship in G-7 countries and emerging markets. Energy Economics, 25 33 37.

15. SRWE, 2007 Statistical Review of World Energy. The British Petroleum Company, 2007.

16. M. M. Uner, N. Kose, S. Gokten, P. Okan, 2008 Financial and Economic Factors Affecting the Lignite Prices in Turkey: An Analysis of Soma and Can Lignite. Resources Policy, 33 4 2008, 230 239 .

17. O. Williamson, 2000 The new institutional economics: taking stock, looking ahead. Journal of Economic Literature, 38 (3), 595-613.

Chapter 7

Modeling and Optimization of the Medium-Term Units Commitment of Thermal Power

Shengli Liao, Zhifu Li, Gang Li *, Jiayang Wang and Xinyu Wu

Institute of Hydropower System & Hydroinformatics, Dalian University of Technology, Dalian 116024, China

ABSTRACT

Coal-fired thermal power plants, which represent the largest proportion of China's electric power system, are very sluggish in responding to power system load demands. Thus, a reasonable and feasible scheme for the medium-term optimal commitment of thermal units (MOCTU) can ensure that the generation process runs smoothly and minimizes the start-up and shut-down times of thermal units. In this paper, based on the real-world and practical demands of power dispatch centers in China, a flexible mathematical model for MOCTU that uses equal utilization hours for the installed capacity of all thermal power plants as the optimization goal and that considers the award hours for MOCTU is developed. MOCTU is a unit commitment (UC) problem with characteristics of large-scale, high dimensions and nonlinearity. For optimization, an improved progressive optimality algorithm (IPOA) offering the advantages of POA is adopted to overcome the drawback of POA of easily falling into the local optima. In the optimization process, strategies of system operating capacity equalization and single station operating peak combination are introduced to move the target solution from the boundary constraints along the target isopleths into the feasible solution's interior to guarantee the global optima. The results of a case study consisting of nine thermal power plants with 27 units show that the presented algorithm can obtain an optimal solution and is competent in solving the MOCTU with high efficiency and accuracy as well as that the developed simulation model can be applied to practical engineering needs.

KEYWORDS

medium-term unit commitment (UC); progressive optimality algorithm (POA); target isopleths; utilization hours of installed capacity

1. INTRODUCTION

In the past several decades, China's power system has experienced rapid development along with rapid economic growth. Among all electric power sources, thermal power, which mainly consists of coal-fired plants, accounts for the largest proportion of China's power system. By the end of 2012, China's total installed generation capacity reached 1146.76 GW, of which 819.68 GW (approximately 71.5%) was from thermal power, and the total electric energy production reached 4986.5 TWh, with 3925.5 TWh (approximately 78.7%) coming from thermal power. Simultaneously, the utilization hours of the installed capacity for thermal power plants are larger than those of hydro power or wind power. The statistical results of 2012 showed that the annual utilization hours of power plants with an installed capacity over 6 MW are 4982 h for thermal power, 3591 h for hydropower and 1929 h for wind power [1,2,3,4,5]. However, because the preheating and cooling of the stream-turbine requires more time than other resource units, the coal-fired thermal unit is inflexible between start-up and shut-down resulting in a slow response to system load demands. Moreover, the start-up and shut-down of coal-fired thermal units cause high fuel costs [6,7,8,9]. Generally, thermal plants are usually scheduled to meet base load demands to ensure that the generation process runs smoothly and reduce the start-up and shut-down times of thermal units. Furthermore, the different values of power system demand become larger among adjacent days due to various factors, including heavy weather, minor vacation, and transition periods between dry and wet season for hydro power [10,11]. This situation causes remarkable difficulties in preparing the generation schedule for power systems, especially for ones with numerous thermal plants. Consequently, the power dispatch centers of China are in charge of the medium-term optimal commitment of thermal units (MOCTU), which usually span from one month to half a year and take one day as the time interval in calculations to obtain the numbers and times of the start-up and shut-down units for every thermal plant each day before actual short-term and daily-ahead scheduling.

The objective of MOCTU is to find the optimal set of thermal generating units or the boot capacity (sum capacity of units that putting into operation in the first operation of a plant) scheme of a power system to satisfy the system load demand, operational restrictions, reliability constraints, and security requirements in each time period. Thus the MOCTU is a form of unit commitment (UC) problem [8,9,12]. In contrast with short-term UC in power systems, which involves determining a start-up and shut-down schedule of units to meet the required demand during a short-term period such as one day or one week [13,14], MOCTU is responsible for dispatching load to each plant for maintenance scheduling and hydrothermal coordination. The main difference between medium-term and short-term thermal UC optimization is that medium-term optimization takes each plant's operating capacity or unit numbers as decision variables instead of scheduling a single unit by short-term operation [15]. It is well-known that the utilization rate of power generation equipment reflects the generation ability of a thermal plant, and the standby condition of generation equipment reflects the ability of a thermal plant to deal with sudden accidents, and utilization hours of the installed capacity for thermal power plants can reasonably reflect the two abovementioned factors, so that operators can operate and manage plants better. Hence, many power grids take equal accumulated operating hours of installed capacity for all thermal power plants as the primary objective of MOCTU during the selected period. However, due to thermal power's main characteristics, including slow start-up and shut-down, complex constraints, large-scale as well as large complicated demands by power grid, developing and establishing computer simulation and optimization models for thermal UC optimization have been most challenging and complex issues [16,17,18]. Furthermore, to achieve the best possible optimal solution and the largest benefit for thermal power plants and power grid, the model should be simulated as close to reality as possible to suit the practical needs of power dispatch centers, especially in China [19,20,21]. However, according to our retrieval results, there are few studies on MOCTU, especially on the development of simulation models for MOCTU that meet the demands of practical engineering [22,23]. In this paper, a medium-term thermal UC optimization model with equal capacity utilization hours for all thermal power plants is established. The model not only respects the basic principles of electric power dispatch (equality, impartiality and transparency), but also adopts a reward principle that gives extra operation hours to the thermal plants with low energy consumption and dust discharge and high efficiency, namely reward hours.

The work described in this paper is originated from the real, practical demands for power dispatch centers of China to develop a medium-term optimal commitment of thermal units. The aim of this paper is to develop a simulation model for MOCTU to satisfy the needs of practical engineering and present a feasible and effective algorithm to optimize the model. MOCTU is a multi-stage decision problem that involves a highly nonlinear and computationally expensive objective function with a large number of constraints. The progressive optimality algorithm (POA) has been shown to be an effective method for solving multi-stage optimization problems by decomposing a multi-stage decision problem into a series of non-linear programming two-stage problems [24,25,26,27,28], and it is suitable for solving MOCTU. However, it is a difficult task to find feasible solutions for a large-scale MOCTU problem using POA due to its drawback, namely the easily encountered local optimum for complex problems. Therefore, in this paper, an improved progressive optimality algorithm (IPOA) is proposed for a large-scale MOCTU problem with nine thermal plants and 27 units in Yunnan Province of China to validate the effectiveness and practicality of the developed simulation model as well as to improve the quality of optimal solutions of POA. The original optimization problem is first optimized by using a heuristic method and an initial feasible solution is obtained, then POA is adopted to search the optimal solution, and finally two strategies are utilized to adjust the solution from the constraint boundaries into the feasible zone's interior to continue to search for the global solution. Actually, the mid-term boot scheduling plan of thermal plants has been generated by operators in Yunnan power grid (YNPG) using this proposed simulation model and method since 20 May 2013. Over a two-year implementation period, it has given operators more confidence to use it.

This paper is organized as follows: Section 2 gives the formulation of the mathematical model of MOCTU; Section 3 analyzes the characteristics of this problem and presents the IPOA used to solve it; Section 4 applies the model to a real-world scenario with nine thermal plants and 27 units in Yunnan province of China; Finally, Section 5 presents the conclusions.

2. MATHEMATICAL MODEL

2.1. Objective Function

MOCTU is an important and complex task for medium-term scheduling of power grids. On the one hand, reasonable MOCTU results should not only satisfy the complicated operation requirements of thermal plants, but respond to system load changes rapidly. Furthermore, they should meet the requirements of the national energy conservation policy as well as balance the interests between different thermal plants. In actual management and operation of thermal power in China, utilization hours of installed capacity for thermal plants can reflect the utilization rate of power equipment and standby application of thermal plants. Thus, controlling the utilization hours of installed capacity for thermal plants has always been the goal that needs to be considered in MOCTU for many power grids. Based on the abovementioned information, taking the equal utilization hours of installed capacity for thermal plants as the optimization goal, and at the same time, the concept of reward-hours is introduced in the model, though which power grids can reward stations that with small coal consumption flexibly according to energy saving principles. Assuming the total number of plants participating in the calculation is *m*, the objective function can be expressed as follows:

$$h_i = h_j \ \ (i \neq j) \tag{1}$$

where h_i and h_j are the accumulated operating hours of installed capacity for plants *i*, *j*, respectively, in the selected period.

Although the above objective function (Equation (1)) is conceptually clear, it is very difficult to solve directly because the objective formulation is related to all plants and the target result cannot be obtained by computing a single mathematical expression. Considering the fact that variance can be used to describe the dispersion degree of data series, the original objective function is replaced by minimizing the variance of vector h=[h1, h2,···,hi,···,hm], which is composed of plants' capacity utilization hours. The replacement function can be represented as:

$$\min f = \frac{1}{m}\sum_{i=1}^{m}(h_i - \overline{h})^2 \tag{2}$$

$$\overline{h} = \frac{1}{m}\sum_{i=1}^{m} h_i \tag{3}$$

$$h_i = h_i^P + h_i^T - h_i^G \tag{4}$$

$$h_i^T = \frac{1}{N_i} \times \overline{C}_i \times T \times 24 \times r_i \tag{5}$$

$$\overline{C}_i = \frac{1}{T}\sum_{t=1}^{T} C_i^t \tag{6}$$

where the following notations are used:

i, j	plant i and plant j, $i, j = 1, 2, \cdots m$;
T	Number of time steps: $t = 1, 2, \cdots, T$;
h_i	Accumulated operating hours of installed capacity of plant i in the selected period;
h_i^P	Operating hours of installed capacity of plant i accumulated in pre-calculation period;
h_i^T	Operating hours of installed capacity of plant i accumulated in the calculation period;
h_i^G	Extra award hours assigned to plant i;
N_i	Installed capacity of plant i;
$\overline{C}_i$	Average operating capacity of plant i in the calculation period;
r_i	Given load factor of plant i;
C_i^t	Operating capacity of plant i in period t. It is also the decision variable to be solved.

In this model, the values of hPi and hGi are given and do not require special treatment while solving the problem. It should be noted that, under the condition of hPi being unequal to hGi, to equalize hi, hTi has to be made unequal. Thus, it realizes the differentiation of capacity utilization hours of all thermal power plants in the calculation and ensures the flexibility of the model.

2.2. Constraints

For MOCTU, each individual thermal plant is subjected to its own set of constraints, while the power system is subjected to system power balance constraints. Specifically, we consider the following constraints:

- **Unit number constraints:**

$$\underline{n}_i^t \leqslant n_i^t \leqslant \overline{n}_i^t \tag{7}$$

where $\underline{n}_i^t$ and $\overline{n}_i^t$ are the minimum and maximum unit number of plant *i* in period *t*, respectively; nti is the active unit number of plant *i*in period *t*.

- **System power balance constraints:**

$$C_S^t \times \underline{r} \times \Delta h_t \leqslant P_D^t \times \Delta h_t \leqslant C_S^t \times \overline{r} \times \Delta h_t \tag{8}$$

where CtS is the operating capacity of the system in period t in MW; Δht is the number of hours in period t in *h*, and here Δht=24; PtD is the load demand of the system in period t in MW; r− and r¯ are, respectively, the minimum and maximum load factors of thermal power, which are calculated from the long-term real-world operation of power grid, r−,r¯∈[0,1], and rm=(r−+r¯)/2 is recorded as the mean-value of the load factor of thermal power.

- **Peak and valley duration constraints:**

The medium-term thermal UC optimization only dispatches system load to plants, and there is no need to distribute load among units. Therefore, we transform minimum up- and down-time constraints into minimum peak and valley time constraints:

$$\begin{cases} T_i^{\mathrm{up}} \geqslant \underline{T}_i^{\mathrm{up}} \\ T_i^{\mathrm{down}} \geqslant \underline{T}_i^{\mathrm{down}} \end{cases} \tag{9}$$

where Tupi is the minimum duration time during the peak in the operating capacity process of plant i in calculation period; T−upi is the limit value of the minimum duration time in the peak in the operating capacity process of plant *i* in the calculation period; Tdowni is the minimum duration time in the valley period in the operating capacity process of plant *i*; T−downi is the limit value of the minimum duration time in the valley period in the operating capacity process of plant *i*.

- **Boundary condition constraints:**

The plants' generating scheme is not only related to the calculation periods, but also relevant to the boot capacity scheme in previous periods, namely pre-calculation periods, where the start-up mode has been determined, because the boot capacity scheme in the pre-calculation period has an impact on the generating scheme in current periods. Therefore, the target start-up mode in calculation periods should link up with existing *p* days start-up mode in calculation periods. The relationship among all periods is shown in Figure 1.

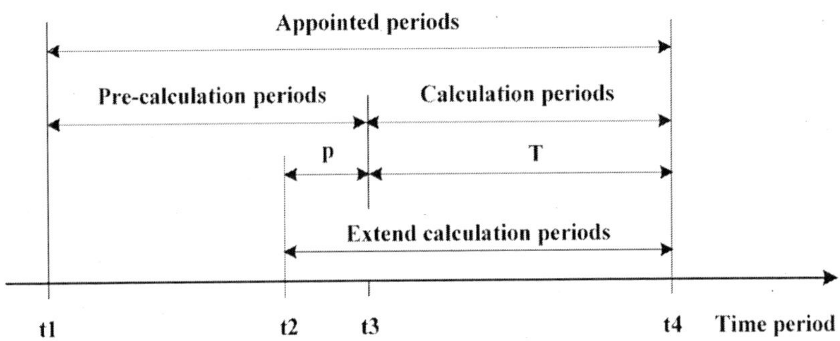

Figure 1. Sketch map of the time relationship.

Note that the actual unit number and operating capacity of a plant are variables to be calculated, and the others are given input data.

3. MODEL SOLUTION

3.1. Solution Approach

MOCTU is a form of the UC problem with a large number of constraints, including unit number constraints (Constraint Group 1), system power balance constraints (Constraint Group 2), peak and valley duration constraints (Constraint Group 3), and boundary condition constraints (Constraint Group 4). It is obvious that Constraint Groups 1 and 2 are single-period constraints that can be treated to reduce the search space. Although Constraint Groups 3 and 4 are multi-period constraints, Constraint Group 4 can be satisfied when Constraint Group 3 is satisfied in extended calculation periods. On the other hand, MOCTU is a high dimensional problem to be optimized. For a power grid with two thermal plants, each having three units. For one plant with three units, it has (3 + 1) kinds of combination in one period, and for two plants, it reaches $(3 + 1)^2$, so the total number of unit combinations will reach $[(3 + 1)^2]^{30} \approx 1.329 \times 10^{36}$ for a horizon of one month with 30 periods, and the overhead of optimization will be computationally expensive. Considering the complex constraints, especially Constraint Group 3, the solving process can be divided into two procedures:

Procedure I: Obtaining the feasible solution space S=[S1,S2,⋯,St,⋯,ST], where St represents all of the possible unit-commit combinations that satisfy Constraint Groups 1 and 2 in period *t*. First, combine all units of all plants to obtain all possible unit-commit combinations in each period. Second, obtain the solution space S by filtering out the combinations that cannot satisfy Constraint Groups 1 and 2 period-by-period. The solution for the UC scheme in the calculation period can be acquired from the unit-commit combinations in S. The above process is relatively simple.

Procedure II: Search for the global optimal solution that can meet Constraint Group 3. Because the solution in S from *Procedure I*already satisfies Constraint Groups 1 and 2, and as well as Constraint Group 4 being automatically satisfied after satisfying Constraint Group 3, the aim of this search process is to find the solution that satisfies Constraint Group 3. The following discussion will introduce how to find the optimal scheduling that meets Constraint Group 3 in the solution space S. First, generate an initial feasible solution in Sthat satisfies Constraint Group 3 by using heuristic method. Second, produce the global optimal solution by the POA. However, POA is sensitive to initial trajectories and sometimes cannot guarantee a global optima, which is the reason that restricts its application. Thus, POA easily converges to a local optimum while being directly used for MOCTU, which is a strong constrained optimization problem. For the medium-term thermal UC optimization, a local optimum means that the system power reaches the upper or lower limit boundary constraints or that the duration periods of peak and valley in the operating capacity process are equal to the limit value of the minimum duration periods. In this case, it is very difficult to further optimize the result. Therefore, an improved POA combining POA with two adjustment strategies is presented to overcome the mentioned demerits.

3.2. Initial Feasible Solution Generation

As *Procedure I* has achieved the solution space S that meets Constraint Groups 1 and 2, the initial feasible solution can be acquired by searching the solution space S by heuristic method only when the operating load rate approximates to the given target load rate that obtained from long-term schedule of power dispatch centers in each period. The feasible region in the latter period will be sharply contracted when the generating scheme in the previous period is determined and the duration period constraints (Constraint Group 3) are considered. To avoid the infeasible region, this paper suggests rm as the target load rate. Before searching, sort the generating schemes in St in every period according to the ascending order of the index value |rk−rm|. Where rk represents the operating load rate of the *k*th generating scheme in St, and its formula is as follows:

$$r_k = \sum_{i=1}^{m} C_i^k / \sum_{i=1}^{m} N_i \qquad (10)$$

The generating scheme S′t, which is sorted according to the ascending order of the index value |rk−rm| in every period, can be obtained by using Equation (10). Then, the feasible solution can be obtained from the sorted solution spaceS′=[S′1,S′2,⋯,S′t,⋯,S′T] by the heuristic method. The detailed procedure is as follows:

Step 1: Construct an integer array *ks* of length *T*. Set *ks*[1] = 1 and *t* = 1. Make sure that the combination of boot capacity in period 1 is the first element of the solution space S′t.

Step 2: Set *t* = *t* + 1. If *t* > *T*, go to Step 6. Otherwise, set *ks*[1] = 1 and set the combination of boot capacity in period *t* as the first element of the solution space S′t.

Step 3: Verify whether the start-up mode lying in the interval of [0,1] satisfies the constraints of Equation (3). If it does satisfy them, go to Step 2; otherwise, go to Step 4.

Step 4: Set *ks*[*t*] = *ks*[*t*] + 1. If *ks*[*t*] is greater than the number of elements in solution space S′t, go to Step 5. If not, replace the combination of boot capacity in period *t* with the *ks*[*t*]th element of the solution space S′t and go to Step 3.

Step 5: Set $t = t + 1$, and go to Step 1.

Step 6: Output the result.

3.3. Optimization Process of Progressive Optimality Algorithm (POA)

Considering the complex requirements by power grids and the constraints mentioned above, the present optimization problem exhibits multi-stage, large-scale and high dimension characteristics. A suitable solution algorithm is required for solving the MOCTU. Based on Bellman's Principle, the POA, which is proposed by Howson and Sancho for reducing dimensionality difficulties by decomposing a multi-stage decision problem into a series of non-linear programming two-stage problems [29], has been shown to have great advantages over classical optimization methods as one of the most widely used techniques for hydroelectric generator scheduling and water resources problems [30,31]. The advantages of POA over other optimization techniques are that it can decompose a multi-state decision problem into several nonlinear programming sub-problems to reduce the dimensionality. More specifically, the merits of POA have clearly been elaborated in [32], including no need to discretize the state variables, no resolution or linearization of nonlinear objective functions and constraints and minimal storage requirements.

To understand the main principle of POA, a general multi-stage optimization example was given. In this example, a set of state variables x∗0,x∗1,x∗2,⋯,x∗T−1,x∗T is given to determine the optimal solution by minimizing objective functio f(x0,x1,x2,⋯xT), where x0 and xT are given as the initial state values. The algorithm starts with an initial trajectory x0,x01,x02,⋯,x0T−1,xT which is gained in a certain way. Then, this multi-stage decision problem would be decomposed into two problems, each of which minimizesf(xji−1,xji,xji+1) to obtain optimal xj,*i at stage *i* during the *j*th iteration by fixing values of xji−1 and xji+1. In other words, xj0 and xj2are fixed to optimize xj1 to yield xj,*1, as shown in Figure 2a. Then xj1=xj,*1 and xj3 are fixed to optimize xj2 for xj,*2, as shown inFigure 2b, in turn, the *j*th iteration isn't finished until xj,*T−1 is obtained, as shown in Figure 2c,d. Based on the optimized results from the last iteration, another iteration is restarted again. The process isn't over until the difference between the last two iterations meets the predefined precision limit. Figure 2 illustrates the optimization process during the *j*th iteration.

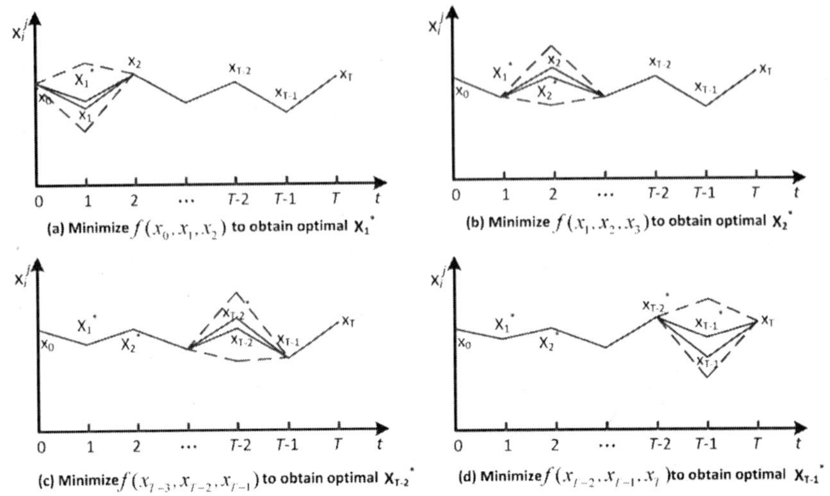

Figure 2. Optimization process of progressive optimality algorithm (POA).

The initial feasible solution obtained in Section 3.3 is used for POA to search for optimal solution. The general process can be mainly described in the following steps:

Step 1: Set the generating scheme to an initial feasible solution generated by the heuristic search and an initial objective function value value. Set $t = 1$ and $k = 1$.

Step 2: Obtain a generating scheme by replacing the generating scheme in period t with the kth element of the solution space St. If the new generating scheme meets Constraint Group 3, go to Step 3. If not, go to Step 4.

Step 3: Calculate the objective function value value' according to Equation (5). If value'<value, adjust the results by using the equalization operation and combination operation and set value'=value.

Step 4: Set $k = k + 1$. If k is greater than the number of elements in solution space St, go to Step 5. If not, go to Step 2.

Step 5: Set $t = t + 1$ and $k = 1$. If $t < T$, go to Step 2. If not, it means one iteration has been performed, then check whether the objective value has been improved over the previous iteration. If so, set the initial solution as the acquired result and go to Step 2. If not, go to Step 6.

Step 6: Output the result.

3.4. Equalization Operation and Combination Operation

Although possessing advantages of quick and strong convergence, POA easily falls into sub-optimal results because each iterative calculation is merely associated with a single period. Unable to make full use of the multi-stage information of the original problem, the sub-optimal results are easily attained at the bounds of constraints. To enhance the searching space and obtain the global optimal result, two strategies are presented including: equalization operation of the system operating capacity and single station boot peak combination operation.

The equalization operation of the system operating capacity is for making the start-up units' capacity equal to the shut-down units' capacity as much as possible by means of adjusting the ratio of the system load demand for each time and operating capacity to approach rm. The equalization operation has two purposes: (1) to make the system generating scheme move from the power constraint boundary into the feasible solution's interior; and (2) to make it more convenient to move and merge the peak of the operating capacity. The equalization process is equivalent to solving the following programming problem:

$$\min f = \sum_{t=1}^{T} \left| C_s^t r_m - P_D^t \right| \quad (11)$$

The constraints are the same as those we have previously mentioned. The approach needs to traverse the generating scheme of each plant in turn and to find whether it satisfies all constraints when performing start-up (shut-down) of one unit in period *t* and shut-down (start-up) of one unit in period t+j (j>0). If it satisfies the constraints, the problem can be optimized with Equation (11) as the target. In the search process, the system's operating capacity remains unchanged, and it can guarantee that the capacity utilization hours of every plant are unchanged. The search process is shown in Figure 3.

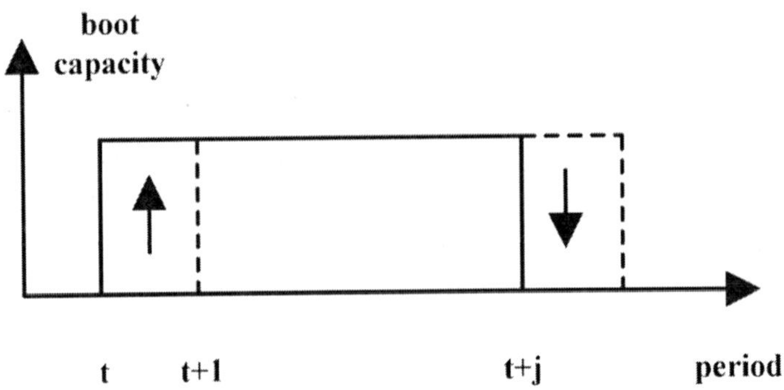

Figure 3. Sketch map of the system's boot capacity equalization.

As the equalization operation makes the start-up unit's capacity equal to the shut-down unit's capacity as much as possible, it provides the peak moving of a single plant's operating capacity some space for adjustment. The single station boot peak combination process searches the single station boot capacity in turn for the peak with the minimum number of duration periods and finds whether it should move and merge it with other peaks. If so, it executes the combination operation. Without changing the objective function value, this method can make an adjustment to the operating capacity process, which requires that the operating durations of peak and valley periods equalize to the demanded minimum duration, and forces it to leave the local optimum. The peak combination operation is shown in Figure 4.

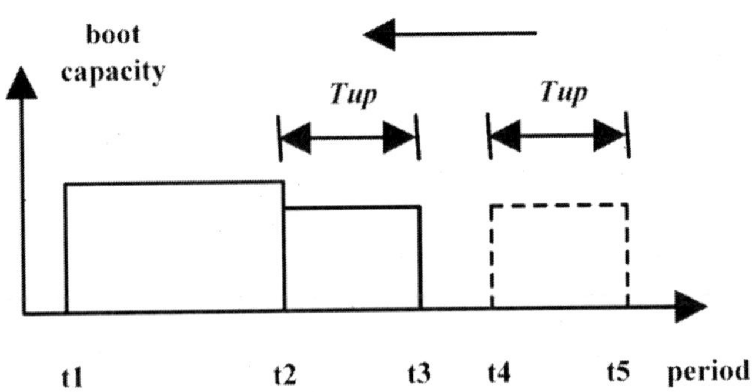

Figure 4. Sketch map of the single station boot peak combination.

From Equations (2)–(6), it can be seen that the installed capacity utilization hours of a thermal power plant is relative to the accumulative total but not the process of the scheme. Utilizing the two operations mentioned above, the results from POA can be converted by making the generating scheme move from the constraint boundary to the feasible region internal area under the condition of the unchanged objective function. As a result, POA acquires its new search space and will not stop iterating until convergence is obtained. The search principle of POA with the two strategies is shown in Figure 5. The lines of f=F0, f=F1 and f=F2 represent the isopleths of the objective function value. According to Equation (5), the objective function value is zero for the best solution x∗, namely f(x∗) = 0. Two main steps are defined in the optimization process. Step α is defined as the optimization process of POA and β presents the adjustment process of equalization operation and merge operation. First, the initial feasible solution x0 converges to the local optimal solution x1 through Step α which cannot be further optimized because x1 lies on the constraint boundary. Secondly, through Step β the local optimal solution x1 can be moved along the target isopleth f=F1 to x2 which is still in the feasible solution internal area, then step α is reutilized and solution x3 is obtained. If x3 still lies on the constraint boundary, then step β is utilized again and the local optimal solution x3 is moved along the target isopleth f=F2 to x4. Based on the optimized results from the adjustment process, another iteration is restarted again until the global optimal solution x∗ is obtained.

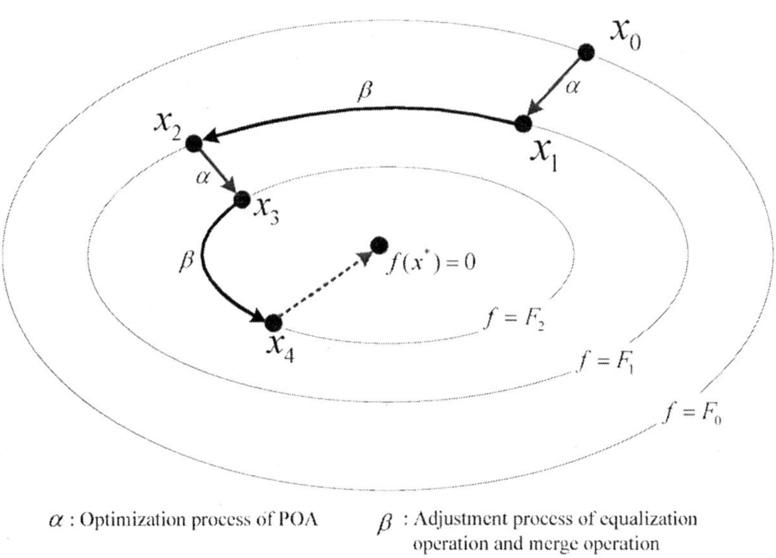

Figure 5. Sketch map of the search principle.

3.5. Solution Architecture

With a large amount of constraints and complex requirements, MOCTU is a challenging optimization problem that involves a highly nonlinear and computationally expensive objective functions. The optimization process is divided into two procedures: generation of an initial feasible solution and optimization of POA. However, POA is sensitive to initial trajectories and easily converges to local optima. Thus any changes from the decision or state variables may have a slight improvement on the solutions. Therefore, IPOA is adopted with the equalization operation and the combination operation to adjust the results acquired from POA. By doing this, it moves the solutions from constraint boundaries into the feasible zone's interior and then continues to search for the optimal solution with POA. The solution flowchart of MOCTU is shown in Figure 6 and the general process for the IPOA procedure can be mainly described in the following steps:

Step 1: Obtain the initial feasible solution by heuristic search and calculate the objective function value;

Step 2: Obtain the optimal results by using POA and calculate the objective function value' (the detailed process is described inSection 3.3);

Step 3: Judge whether the current isopleths of the objective function value are equal to zero, if not, use the two strategies to move the optimal results from the constraint boundary to the feasible region internal area to search again. Otherwise, go to Step 4.

Step 4: Output optimal scheduling.

4. SIMULATION RESULTS

The proposed simulation model and method have been applied to the YNPG in China. Currently, it is used as the primary tool to determine the mid-term boot scheduling of thermal plants by the operators of YNPG. By the end of 2013, the total installed capacity of YNPG had reached 26.1 GW among which hydro power was responsible for 16.7 GW and thermal power was responsible for 8.935 GW. The reasonable scheduling of thermal power is beneficial for optimizing hydropower systems and system security.

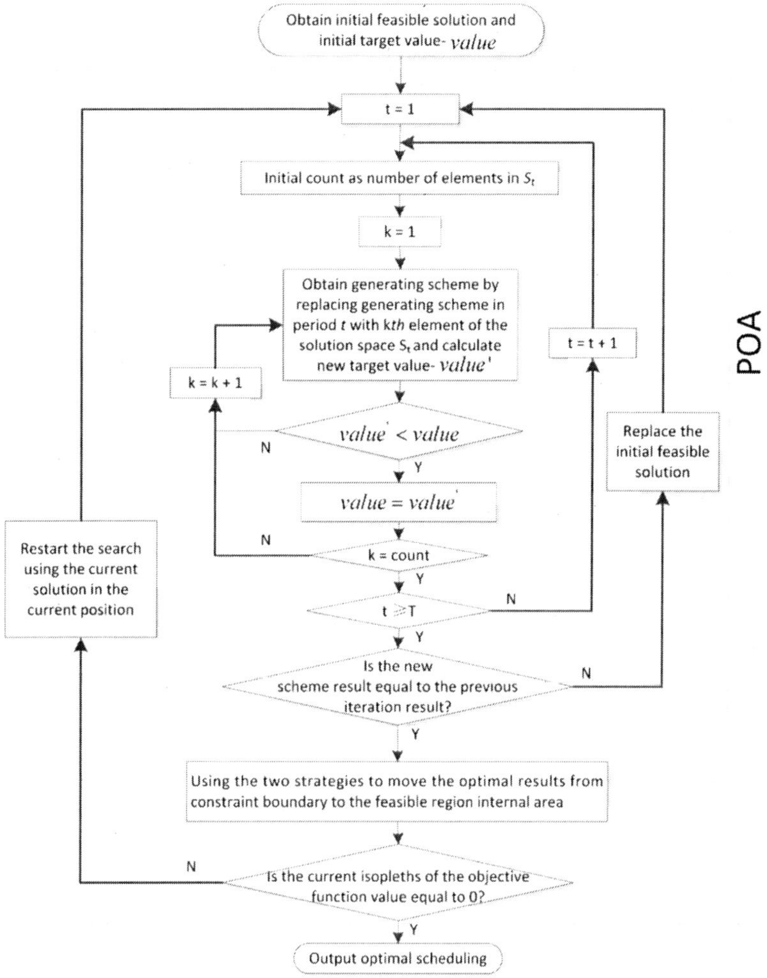

Figure 6. Solution flowchart of medium-term optimal commitment of thermal units (MOCTU) by improved progressive optimality algorithm (IPOA).

The thermal power system in the YNPG consists of nine thermal power plants and 27 units. As mentioned above in Section 3.3, it is a highly dimensional and complex problem to optimize. Therefore, it is a substantial challenge to determine the medium-term operating policies for these thermal plants. The dispatch center in the YNPG is in charge of these thermal power systems. Table 1 lists the basic data of these plants. IPOA has been implemented in Java on a PC with an Intel® Core™2 Duo CPU, operating at 2.93 GHz, with 2 GB of memory. A real scheduling for 2013 in the YNPG is used to test the validity and computational efficiency of the proposed method.

October, a typical month and the beginning of the dry season, was selected to demonstrate the actual availability of the simulation model as well as the practicality and efficiency of the method. In October, the hydro-power system comes to its lowest output and the demand on the thermal system increases. As a result, the operating scheduling of thermal power plants will obviously change. The simulation results and rationality of the algorithm will be presented in the paper. According to the real-world operating experience and users' actual demands, the parameters are set as follows:

1) Given load rate of plant i, ri=0.8;

2) Minimum load rate of thermal power, r−=0.7;

3) Maximum load rate of thermal power, r¯=0.9;

4) Minimum duration periods of peak in the operating capacity process of plant i during the calculation period, Tupi=7d;

5) Minimum duration periods of valley in the operating capacity process of plant i during the calculation period, Tdowni=3d;

6) Maximum unit number of plant *k* in period *t*, set nti¯¯¯ as the number of installed units

7) Minimum unit number of plant *i* in period *t*, nti−−=1(i=A~H), except nti−−=0 while i=I;

8) Actual capacity utilization hours of plant *i* in the pre-calculation period, hPi=0;

9) Extra award hours of plant *i*, hGi=0;

10) System load demands in each period and the first ten days' boot process (p=10d) are all given.

Table 1. Thermal power system in Yunnan power grid (YNPG).

Thermal plant	Units (number × capacity, MW)	Capacity (MW)
A	4 × 600	2400
B	6 × 300	1800
C	4 × 300	1200
D	2 × 200 + 2 × 300	1000
E	2 × 300	600
F	2 × 300	600
G	2 × 300	600
H	2 × 300	600
I	1 × 135	135

Comparison of each power plant's capacity utilization hours calculated by IPOA in October is shown in Table 2.

Table 2. Comparison of each power plant's capacity utilization hours in October (h).

Items	Heuristic search	POA	First adjustment	IPOA
Plant A	547.2	422.4	422.4	422.4
Plant B	297.6	416.0	416.0	422.4
Plant C	369.6	422.4	422.4	422.4
Plant D	240.0	422.4	422.4	422.4
Plant E	489.6	422.4	422.4	422.4
Plant F	489.6	432.0	432.0	422.4
Plant G	499.2	422.4	422.4	422.4
Plant H	528.0	432.0	432.0	422.4
Plant I	422.4	422.4	422.4	422.4
Average value	431.5	423.8	423.8	422.4
Max-min difference	307.2	16.0	16.0	0
Objective value (h^2)	10283	23.0	23.0	0

First adjustment means the results from the first equalization operation and combination operation.

Figure 7 and Figure 8, respectively, show system energy balance maps and boot modes of plant *F* by different methods (the horizontal axis represents the time period). As can be seen from Figure 7, the results of the simulation model match the actual characteristics of thermal power system in October that the load demand on the thermal system is increasing. It demonstrates that the proposed simulation model for MOCTU is very practical and can satisfy the actual project requirements.

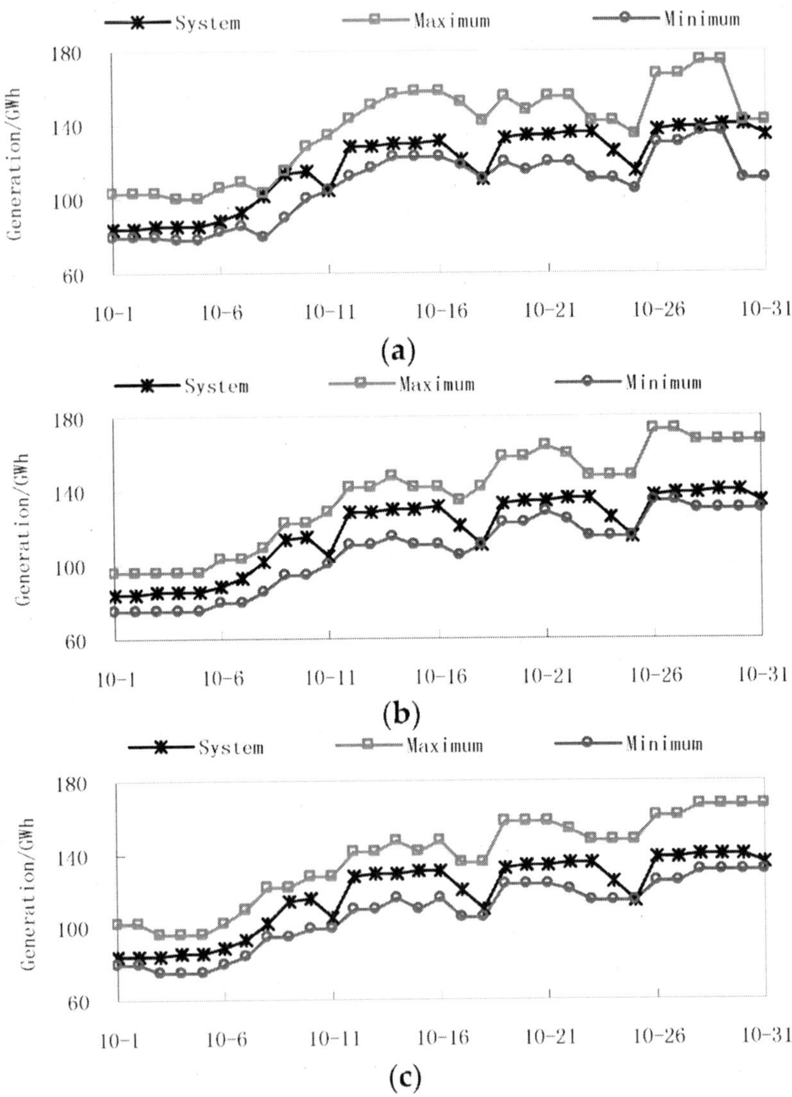

Figure 7. Energy balance maps by three methods: (**a**) POA; (**b**) first adjustment; (**c**) IPOA.

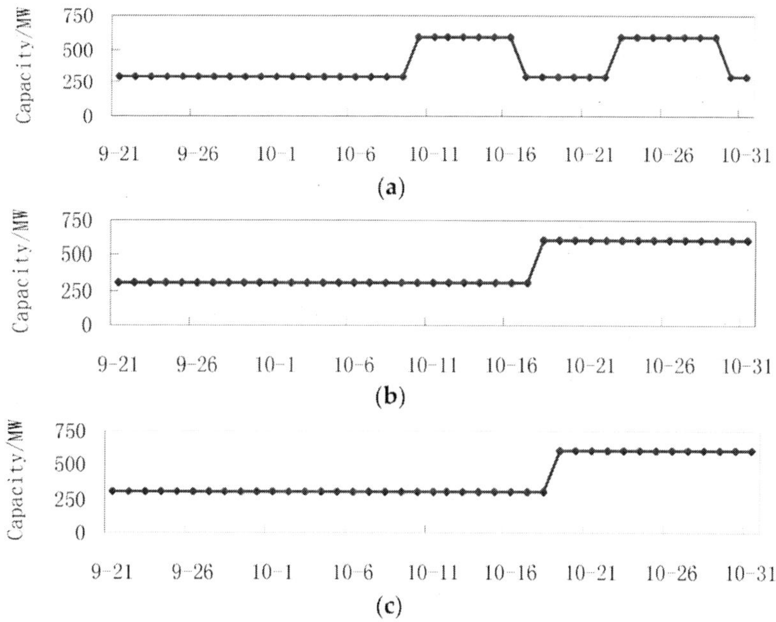

Figure 8. Boot capacity of Plant F for three methods: (**a**) POA; (**b**) first adjustment; (**c**) IPOA.

The calculation results on the YNPG are shown in Table 2. IPOA is compared with heuristics and POA. Max-min difference and objective value are taken as the measurement index. When the initial solution is obtained by heuristics, most plants' capacity utilization hours are different. As shown in Table 2, calculated by heuristic and POA in turn, the max-min differences are 307.2 h and 16 h, and the objective values are 10,283.0 h^2 and 23.0 h^2. Although the results have been significantly improved, the optimal solution is still not found. The reason is that the results obtained from POA run into local optima. The results by POA in Figure 7a illustrate that it has reached the system's power constraint boundary in many periods (10-08, 10-09, 10-30) and Figure 8a shows that two peak values of Plant *F*'s boot mode reach the duration periods of peak in the operating capacity constraint, and it cannot be optimized any further by POA. Table 2, Figure 7b and Figure 8b show that the equalization operation and combination operation can change the structure of solutions and make the solution leave the constraint boundary under the condition of the unchanged objective function value (23.0 h^2). After the first combination operation, the boot process of Plant *F* combines two peaks (Figure 8a) into one peak (Figure 8b) and makes the duration periods of the peak change from 7 d (T^{up} = 7) to 14 d, leaving the constraint boundary. After applying IPOA, the second unit's boot time of Plant *F* is adjusted from 18 October (Figure 8b) to 19 October (Figure 8c), and the capacity utilization hours change from 432 h to 422.4 h.

At the same time, Table 2 shows that it has found the optimal results of the problem, which indicates that the strategies of equalization operation and combination operation can provide a new round of the search for the optimization space and finding the global optimum. At this moment, the optimal solution is obtained, as shown in Table 2, which manifests that the equalization operation and combination operation provide the search space for a new iteration to find the optimal solution. The generation requirements for the nine plants in October are shown in Figure 9 and the details of the results are listed in Table 3.

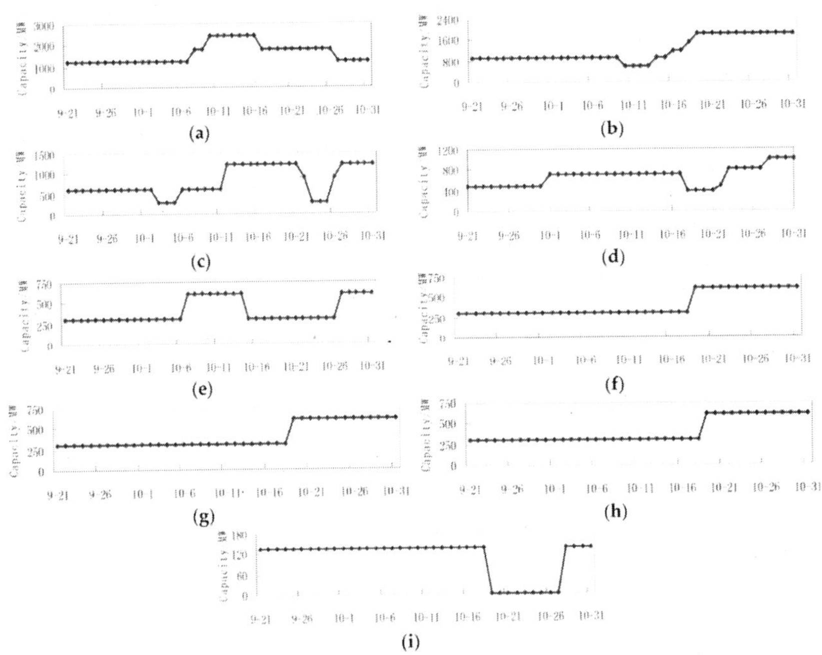

Figure 9. Results of thermal plants: (**a**) Plant A; (**b**) Plant B; (**c**) Plant C; (**d**) Plant D; (**e**) Plant E; (**f**) Plant F; (**g**) Plant G; (**h**) Plant H; (**i**) Plant I.

From Table 3, it can be seen that the boot capacity result of each plant can meet all constraints. In Plant B, for example, the days of the first valley in plant output are 4 (from 10th to 13th) which is larger than the minimum duration of valley in plant output Tdowni=3d, while the days of first peak in plant output are 13 (from 19th to 31st) which is larger than minimum duration of peak in plant outputTupi=7d. Figure 9 lists the boot processes for each plant achieved by IPOA and all of the boot processes are satisfied Constraint Groups 3 and 4, which demonstrate the effectiveness of the proposed algorithm.

For comparison and illustration the impact of extra award hours, namely hGi, attach extra award hours for Plant C, D and F, and the award hours are 30, 20 and 10, respectively. The other conditions are unchanged. The optimization results are shown in Table 4. Because values of installed capacity utilization hours of each plant is discrete in calculation periods, the objective function value corresponding to optimal solutions is not always zero, and as a result, the objective value becomes 0.057 h^2. Moreover, installed capacity utilization hours of Plants C, D and F are more 29.52, 19.36 and 9.68 h than other plants, respectively, and the calculation precision meets the demand of practical engineering.

5. CONCLUSIONS

According to the real and practical demands of power dispatch centers of China, a simulation model for medium-term thermal UC is presented and solved by using an efficient solution named IPOA. As a new contribution to thermal power UC for power grids, the simulation model gives plenty of considerations to load demands, plant characteristics, practical engineering needs and other important factors. IPOA, which is developed to improve the performance of POA, is another contribution. The major advantage of IPOA is that strategies of operating capacity equalization and single station operating peak combination are performed to move the feasible solution away from constraint boundaries, along target isopleths, into the feasible zone's interior, in order to provide the search space for a new iteration to overcome the drawback of easily converging to local optima of POA. Basically, the mid-term boot scheduling plan of thermal plants has been generated by operators in YNPG using the proposed simulation model and method since 20 May 2013. Two-year implementation and the results of thermal power plants with 27 units in YNPG demonstrate that the proposed model has made a closer step from theoretical developments to real-world implementations, and that IPOA is an effective method for solving the MOCTU problem.

Table 3. Results of generation requirements for plants in October (MW).

Date	Plant A	Plant B	Plant C	Plant D	Plant E	Plant F	Plant G	Plant H	Plant I
1st	1200	900	600	700	300	300	300	300	135
2nd	1200	900	600	700	300	300	300	300	135
3rd	1200	900	**300**	700	300	300	300	300	135
4th	1200	900	300	700	300	300	300	300	135
5th	1200	900	300	700	300	300	300	300	135
6th	1200	900	**600**	700	300	300	300	300	135
7th	1200	900	600	700	**600**	300	300	300	135
8th	**1800**	900	600	700	600	300	300	300	135
9th	1800	900	600	700	600	300	300	300	135
10th	**2400**	**600**	600	700	600	300	300	300	135
11th	2400	600	600	700	600	300	300	300	135
12th	2400	600	**1200**	700	600	300	300	300	135
13th	2400	600	1200	700	600	300	300	300	135
14th	2400	**900**	1200	700	600	300	300	300	135
15th	2400	900	1200	700	**300**	300	300	300	135
16th	2400	**1200**	1200	700	300	300	300	300	135
17th	**1800**	1200	1200	700	300	300	300	300	135
18th	1800	**1500**	1200	**400**	300	300	300	300	135
19th	1800	**1800**	1200	400	300	**600**	**600**	**600**	**0**
20th	1800	1800	1200	400	300	600	600	600	0
21st	1800	1800	1200	400	300	600	600	600	0
22nd	1800	1800	**900**	**500**	300	600	600	600	0
23rd	1800	1800	**300**	**800**	300	600	600	600	0
24th	1800	1800	300	800	300	600	600	600	0
25th	1800	1800	300	800	300	600	600	600	0
26th	1800	1800	**900**	800	300	600	600	600	0
27th	**1200**	1800	**1200**	800	**600**	600	600	600	0
28th	1200	1800	1200	**1000**	600	600	600	600	**135**
29th	1200	1800	1200	1000	600	600	600	600	135
30th	1200	1800	1200	1000	600	600	600	600	135
31st	1200	1800	1200	1000	600	600	600	600	135

Table 4. Comparison of capacity utilization hours of each plant (h).

Items	h_i^G	Heuristic Search		IPOA	
		$h_i^P + h_i^T$	h_i	$h_i^P + h_i^T$	h_i
Plant A	0	240.00	240.00	422.40	422.40
Plant B	0	297.60	297.60	422.40	422.40
Plant C	30	399.12	369.12	451.92	421.92
Plant D	20	508.96	488.96	441.76	421.76
Plant E	0	422.40	422.40	422.40	422.40
Plant F	10	499.28	489.28	432.08	422.08
Plant G	0	547.20	547.20	422.40	422.40
Plant H	0	499.20	499.20	422.40	422.40
Plant I	0	528.00	528.00	422.40	422.40
Average value	-	-	431.50	-	422.24
Max-min difference	-	-	307.20	-	0.64
Objective value (h^2)	-	-	10,277.489	-	0.057

ACKNOWLEDGMENTS

This study is supported by the National Natural Science Foundation of China (No. 51209029), the Fundamental Research Funds for the Central Universities (20120041120002) and the Fundamental Research Funds for the Central Universities (No. DUT14QY15).

AUTHOR CONTRIBUTIONS

Shengli Liao and Gang Li devised the experimental strategy and carried out this experiment. Shengli Liao and Zhifu Li wrote the manuscript and contributed to the revisions. Jiayang Wang and Xinyu Wu contributed to the experiment and analysis of the data.

CONFLICTS OF INTEREST

The authors declare no conflict of interest.

REFERENCES

1. China Electric Power Yearbook Editorial Board. *China Electric Power Yearbook*; China Electric Power Press: Beijing, China, 2011.

2. China Electric Power Yearbook Editorial Board. *China Electric Power Yearbook*; China Electric Power Press: Beijing, China, 2012.

3. China Electric Power Yearbook Editorial Board. *China Electric Power Yearbook*; China Electric Power Press: Beijing, China, 2013.

4. Ministry of Planning and Statistics. *National Power Industry Statistics Express*; China Electricity Council: Beijing, China, 2013.

5. Liu, L.; Zong, H.; Zhao, E.; Chen, C.; Wang, J. Can China realize its carbon emission reduction goal in 2020: From the perspective of thermal power development. *Appl. Energy* **2014**, *124*, 199–212.

6. Mario, C.; Antonio, M.; Giuseppina, N. A Dynamic Fuzzy Controller to Meet Thermal Comfort by Using Neural Network Forecasted Parameters as the Input. *Energies* **2014**, *7*, 4727–4756.

7. Padhy, N.P. Unit commitment—A bibliographical survey. *IEEE Trans. Power Syst.* **2004**, *19*, 1196–1205.

8. Niknam, T.; Khodaei, A.; Fallahi, F. A new decomposition approach for the thermal unit commitment problem. *Appl. Energy***2009**, *86*, 1667–1674.

9. Senjyu, T.; Shimabukuro, K.; Uezato, K.; Funabashi, T. A fast technique for unit commitment problem by extended priority list.*IEEE Trans. Power Syst.* **2003**, *18*, 882–888.

10. Baldwin, C.J.; Dale, K.M.; Dittrich, R.F. A study of the economic shutdown of generating units in daily dispatch. *Trans. Am. Inst. Electr. Eng.* **1959**, *78*, 1272–1284.

11. Iguchi, M.; Yamashiro, S. An efficient scheduling method for weekly hydro-thermal unit commitment. In Proceedings of the 2002 IEEE Region 10 Conference on Computers, Communications, Control and Power Engineering, Beijing, China, 28–31 October 2002; Volume 3, pp. 1772–1777.

12. Dudek, G. Genetic algorithm with integer representation of unit start-up and shut-down times for the unit commitment problem.*Eur. Trans. Electr. Power* **2007**, *17*, 500–511.

13. Tong, S.K.; Shahidehpour, S.M.; Ouyang, Z. A heuristic short-term unit commitment. *IEEE Trans. Power Syst.* **1991**, *6*, 1210–1216.

14. Hyeon, G.P.; Jae, K.L.; Kang, Y.C.; Jong, K.P. Unit commitment considering interruptible load for power system operation with wind power. *Energies* **2014**, *7*, 4281–4299.

15. Liao, S.L.; Cheng, C.T.; Wang, J.; Feng, Z.K. A hybrid search algorithm for midterm optimal scheduling of thermal power plants. *Math. Probl. Eng.* **2015**, *2015*.

16. Venkata, S.P.; Istvan, E.; Kurt, R.; Jan, D. A stochastic model for the optimal operation of a wind-thermal power system. *IEEE Trans. Power Syst.* **2009**, *24*, 940–949.

17. Saadawi, M.M.; Tantawi, M.A.; Tawfik, E. A fuzzy optimization-based approach to large scale thermal unit commitment.*Electr. Power Syst. Res.* **2004**, *72*, 245–252.

18. Dang, C.Y.; Li, M.Q. A floating point genetic algorithm for solving the unit commitment problem. *Eur. J. Oper. Res.* **2007**, *181*, 1670–1395.

19. Simonovic, S.P. Reservoir systems analysis: Closing gap between theory and practice. *J. Water Resour. Plan. Manag.* **1992**, *118*, 262–280.

20. Yang, G.; James, M.C.; Ming, N.; Rui, B. Economic modeling of compressed air energy storage. *Energies* **2013**, *6*, 2221–2241.

21. Rogelio, P.M.; Juan, A.M.; Miguel, J.P.; Lourdes, A.B.; Juan, M.M.S. A novel modeling of molten-salt heat storage systems in thermal solar power plants. *Energies* **2014**, *7*, 6721–6740.

22. Wei, S.Y.; Xu, F.; Min, Y. Study and modeling on maintenance strategy for a thermal power plant in the new market environment. In Proceedings of the IEEE International Conference on Electric Utility Deregulation, Restructuring and Power Technologies, Hong Kong, China, 5–8 April 2004; pp. 200–204.

23. Janusz, B.; Andrzej, O. Modelling of thermal power plants reliability. In Proceedings of the International Conference on Power Engineering, Energy and Electrical Drives, Istanbul, Turkey, 13–17 May 2013; pp. 1038–1043.

24. Turgeon, A. Optimal short-term hydro scheduling from the principle of progressive optimality. *Water Resour. Res.* **1981**, *17*, 481–486.

25. Nanda, J.; Bijwe, P.R.; Kothari, D.P. Application of progressive optimality algorithm to optimal hydrothermal scheduling considering deterministic and stochastic data. *Int. J. Electr. Power Energy Syst.* **1986**, *8*, 61–64.

26. Nanda, J.; Bijwe, P.R. Optimal hydrothermal scheduling with cascaded plants using progressive optimality algorithm. *IEEE Trans. Power Syst.* **1981**, *100*, 2093–2099.

27. Lucas, N.J.D.; Perera, P.J. Short-term hydroelectric scheduling using the progressive optimality algorithm. *Water Resour. Res.***1985**, *21*, 1456–1458.

28. Cheng, C.T.; Shen, J.J.; Wu, X.Y.; Chau, K.W. Short-term hydro scheduling with discrepant objectives using multi-step progressive optimality algorithm. *J. Am. Water Resour. Assoc.* **2012**, *48*, 464–479.

29. Labadie, J.W. Optimal operation of multireservoir systems: State-of-the-art review. *J. Water Resour. Plan. Manag.* **2004**, *130*, 93–111.

30. Howson, H.R.; Sancho, N.G. A new algorithm for the solution of multi-stage dynamic programming problems. *Math. Program.***1975**, *8*, 104–116.

31. Marino, M.A.; Loaiciga, H.A. Dynamic model for multireservoir operation. *Water Resour. Res.* **1985**, *21*, 619–630.

32. Marino, M.A.; Loaiciga, H.A. Quadratic model for reservoir management: Application to the Central Valley Project. *Water Resour. Res.* **1985**, *21*, 631–641.

Chapter 8

Oxy–Fuel Combustion in the Lab–Scale and Large–Scale Fuel–Fired Furnaces for Thermal Power Generations

Audai Hussein Al-Abbas[1] and Jamal Naser[2]

[1] Foundation of Technical Education, Al-Musaib Technical College, Babylon, Iraq

[2] Faculty of engineering and Industrial Sciences, Swinburne University of Technology, Hawthorn, Victoria, Australia

1. INTRODUCTION

Recently, the environmental and health threat from anthropogenic emissions of greenhouse gases (GHG) of power plants has been considered as one of the main reasons for global climate change [1]. The undesirable increase in global temperature is very likely because of increase the concentrations of these syngas in the atmosphere. The most important resource of these anthropogenic GHG emissions in the atmosphere is carbon dioxide emissions. At present, fossil fuels provide approximately 85% of the world's demand of electric energy [2]. Many modern technologies in the electricity generation sector have been developed as sources of new and renewable energies. These new technologies include solar energy, wind energy, geothermal energy, and hydro energy. While these sources of renewable energy are often seen as having zero greenhouse gas emissions, the use of such technologies can be problematic. Firstly, sources of renewable energy are often still under development. Therefore, there can be a higher cost involved in their installation and in other related technical requirements. Secondly, the sudden switching of these energy sources (zero emission) has caused serious problems with the infrastructure of energy supply and global economy [3]. In order to reduce the problem and obey the new environmental and political legislation against global warming, it is necessary to find an appropriate solution to cut pollution which is with cost-effective, from the energy sources. The most effective technique, which can achieve a high level of reduction in GHG emission to atmospheric zone, is to capture carbon dioxide from the conventional power generations. At present, several organizations, energy research centres, companies, and universities, particularly in developed countries, are working to develop these conventional power plants in order to make them more environmentally friendly, with near-zero emissions sources. This chapter continues on different CO_2 capture technologies such as pre-combustion capture, post-combustion capture, and oxy-fuel combustion capture. The developments on oxy-fuel combustion technology with different scales of furnaces in terms of experimental investigations and theoretical modelling are discussed. The fundamentals and operating conditions of oxy-fuel-fired power plants are reviewed due to the importance of these conditions on the flame stability and coal combustion behaviour relative to those of conventional combustion. The effects of particular factors and parameters on the oxy-fuel combustion characteristics and boiler performance are reported. Finally, the chapter closes with a comprehensive CFD modelling study on the lab-scale and large-scale furnaces under air-fired and oxy-fuel combustion conditions.

2. DIFFERENT CO_2 CAPTURE TECHNOLOGIES

In order to understand the technologies that are used for CO_2 capture in the conventional power plants, it is important to understand the systems of leading technology for these power plants. The most popular leading technology systems are as follow: Integrated Gasification Combined Cycle (IGCC), Natural Gas Combined Cycle (NGCC) and Pulverised Fuel (PF) combustion steam cycles (some references called Pulverised Coal (PC)). As previously mentioned, large amounts of CO_2 emissions and other gases such as nitrogen oxides (NO_x) and sulphur oxides (SO_x) are produced by energy production from fossil fuel. Several techniques to capture carbon dioxide are being increasingly developed in order to comply with the new environmental and political legislation against global warming [3, 4]. The three main techniques, which have been developed for CO_2 capture from these different systems of leading technology, are pre-combustion capture, post-combustion capture, and capture of oxy-fuel combustion.

2.1. Pre-Combustion Capture

The pre-combustion capture technique excludes CO_2 from the fuel before the burning process in the combustion chamber. This technique can be achieved by installing special equipment which captures CO_2 between the gasifier and the combined cycle power plant. After gasification of coal or the reforming of natural gas with oxygen, the first step of this process leads to the production of a split stream of carbon monoxide (CO) and combustible gases (mainly hydrogen (H_2)). The second step is to convert CO to CO_2 with steam (synthetic gas with suitable amounts of water vapour) by a process called shift-conversion (CO+H2O -→-yields H2 + CO2CO+H2O →yields H2 + CO2). After that, CO_2 can be separated by using a physical solvent, and finally the CO_2 becomes efficiently available for a sequestration process after it passes through a compression unit. The other remaining parts (mainly hydrogen (H_2)) will be sent to the combined cycle power plant to be used as input fuel for power production [5]. As shown in Figure 1, the clean syngas is supplied to a combined cycle power plant after several treatment processes such as gas cooling (for protecting equipment), particulate removal, and hydrogen sulphide (H_2S) removal. Although this method of CO_2 capture can be considered a good producer of hydrogen for the combined cycle power plant, it has a relatively high level of complexity compared with other CO_2capture techniques [6].

2.2. Post–Combustion Capture

The post-combustion capture technique involves capturing CO_2, as well as reducing particulate matter, SO_x, and NO_x in the combustion flue gases (see Figure 2). This technique requires adding a separation unit after firing systems of the PC or NGCC. Any of the following three separating technologies can achieve the sequestration of CO_2: chemical absorption, low temperature distillation, and gas separation membranes [7].

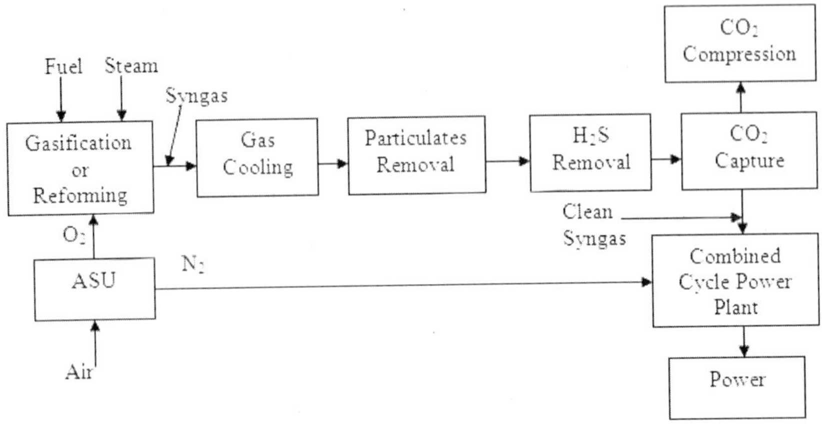

Figure 1. Power plant with pre-combustion capture technology

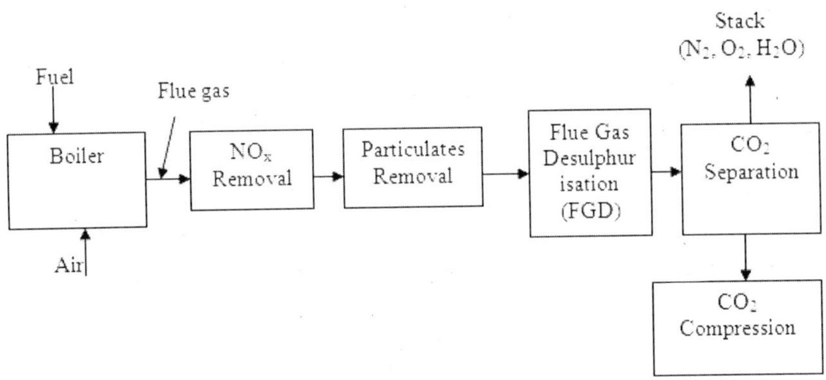

Figure 2. Power plant with Post-combustion capture technology

In the first separating technology, the chemical absorption of chemicals such as monoethanolamine (MEA) is used in an absorption tower to scrub CO_2 from the combustion flue gases. A high temperature level is used to separate CO_2 from the chemical solvent after delivering it to the regenerating tower. After that, a compression process is performed to capture CO_2, as seen in Figure 3. Because of the relatively high temperature and low partial pressure (concentration) of the carbon dioxide to be treated in the flue gases, this method offers a big design challenge for conventional power generation. This means that the chemical absorption process can only provide an economic benefit if it is applied to work at a small scale. On the other hand, this process needs a large amount of energy due to the large size of the main sequestration components, and thereby the energy penalty will introduce a higher operating cost if it is applied for large-scale power plant with CO_2 capture [8, 9]. The second separating technology for post-combustion capture uses gas separation membranes such as solution-diffusion or molecular sieving, which can be used to capture CO_2 by separating it from the flue gases. These membranes can experience some technical problems if applied to the capture of CO_2in flue gases from coal power plants due to the degradation of the absorbent by impurities existing in the flue gas. However, this technology has not yet shown its ability to be applied at a large-scale CO_2capture power plant, and it is still under development. Finally, the third separation technology, low temperature distillation, can be used to capture CO_2 from the flue gas, but this process requires special conditions (above 75 psi pressure, and -75 ^{o}F temperature) to achieve a high purity of CO_2 (about 90% CO_2) in the flue gas. Due to these complicated conditions involved in the separating processes in a power plant, low temperature distillation is not considered an efficient technology for the CO_2capture from power plants.

2.3. Oxy–Fuel Combustion Capture

The oxy-fuel combustion technique captures carbon dioxide from the flue gases of combustion. It is approximately similar to the post-combustion capture technique in terms of separating the CO_2 from the exhaust gases as a final process of sequestration, but it is less chemically complicated. As described earlier, the partial pressure of CO_2 in the flue gas is low in conventional combustion (air-fired combustion), and it needs special treatments for the separation processes. The basic principle of oxy-fuel combustion is to increase the partial pressure of CO_2 in the flue gases in order to make its sequestration and compression process easier and more cost-effective. This technique can be performed by using a mixture of pure oxygen with part of a recycled flue gas (RFG) (mainly CO_2) instead of air in the combustion chamber. In this case of combustion, a high concentration of CO_2 (high partial pressure) can be achieved in the flue gas stream, and therefore the high cost of its capturing processes can be avoided unlike the post-combustion process. The oxy-fuel combustion technique is schematically shown in Figure 3. In Figure 3, the oxy-fuel technique shows that the first removal equipments extract particulates and sulphur dioxide, respectively. The particulate removal can remove the fly ash from the flue gas, while the bottom ash is removed after accumulating at the bottom of the furnace, i.e. at the hopper zone.

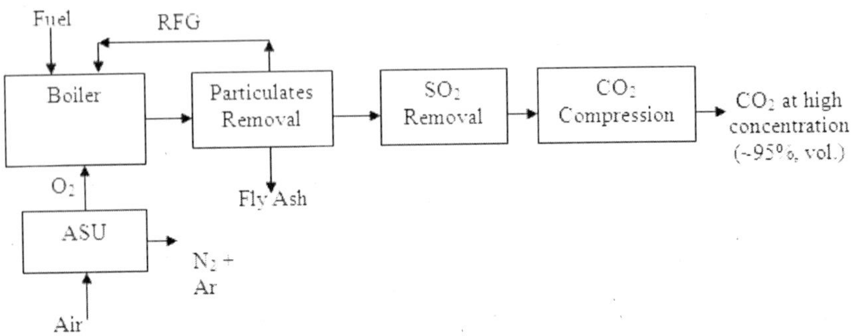

Figure 3. Power plant with the capture of oxy-combustion technology

After removing particulates and condensing water vapour from the flue gases, the concentration of CO_2 will be increased to around 75% vol. under wet basis or to around 95% vol. under dry basis so that it can be transported for permanent storage. The purity of CO_2 concentration is completely dependent on factors such as the purity of the oxygen feed (from ASU), air-leakage, and the excess of the oxygen/fuel (stoichiometry) ratio. As shown in Figure 3, a part of flue gas (around 60-70%) is recycled and mixed with pure oxygen. This process is used to prevent high temperature levels inside the furnace, i.e. to protect the furnace wall. Due to this recirculating of flue gases, the size of the furnace and the sizes of other gaseous removal equipment can be significantly reduced. The recycled gas process of oxy-fuel combustion technique can maintain the same flow field conditions of the burners in the conventional combustion case.

The air separation unit (oxygen production unit) produces two gas streams. The first is an oxygen stream, which supplies the furnace, and the second includes nitrogen and other minor constituent inert gases that are vented to the atmosphere. In the air separation unit (ASU), a large amount of energy is used to separate nitrogen and produce a pure stream of oxygen. Despite the high cost of this process, oxy-fuel combustion is definitely a competitive technique compared with the other CO_2 capture technologies due to the high reduction in NO_x and SO_x emissions besides its high CO_2 capture. However, using pure oxygen and recycled flue gas (RFG) instead of air to burn pulverized coal (PC) in the combustion chamber leads to problems such as many changes in flame temperature levels, species concentrations, and radiation heat transfer problems inside the furnace. These modifications of the combustion characteristics are due to the following reasons: the high specific heat capacity of $CO_2$with respect to nitrogen in conventional combustion, radiative properties of gas mixtures, low oxygen molecular diffusivity in CO_2 compared to N_2 and other transport properties of the gas mixture such as viscosity, thermal diffusivity, gas phase chemistry etc. [1, 3, 10-12].

2.4. Comparison Amongst Different Capture Technologies

In short, all these three CO_2 capture technologies have different outcomes, particularly with regards to reduction of power plant efficiency and in increasing the cost of electricity production. In general, to be successful post-combustion capture requires new developments in the process of chemical absorption of CO_2 in order to adequately reduce energy consumption in the absorption process, but this is very expensive. In contrast, pre-combustion capture is achieved by the conversion of fuel into carbon monoxide (CO) and hydrogen fuel (H_2) in which CO is converted to CO_2 by the shift-conversion process. This CO_2 capture approach can be developed by either physical or chemical absorption processes to avoid any extra complexity in chemical design of power generation. However, both of these processes are very expensive and chemically complicated. Finally, capture of CO_2 by the oxy-fuel combustion technique is less expensive than the other two processes and less complex. It can be carried out by burning the fossil fuel with a mixture of pure oxygen (99.5 vol. %), produced in ASU, and recycled flue gas (RFG). The products of this combustion will be only CO_2 and H_2O in the flue gas. After the condensation process, CO_2 concentrations will be increased to a level more suitable for the separating and compression processes. Although energy consumption for O_2 production, in an air separation unit, is relatively high (about 10% of the net energy of power plant), new technologies for air separation processes and O_2 production are under development in order to reduce the energy penalty for CO_2 capture power plants [13].

Recently, Kanniche *et al.* (2009) [7] comprehensively made a comparison between the leading technology systems for the IGCC, NGCC, and PC power plants with the above-mentioned CO_2 capture technologies. The results showed that the efficiency of the PC power plant when it used post-combustion capture is lower than that of the IGCC power plant using pre-combustion capture. The NGCC and PC power plants obtained the highest efficiencies with post-combustion capture and with oxy-fuel combustion, respectively. Regarding the production costs, the lowest costs occurred when PC used oxy-fuel capture, but costs gradually increased for IGCC use of pre-combustion capture and NGCC use of post-combustion capture. The highest cost per tonne of CO_2 removal was for NGCC using pre-combustion capture, whereas the costs for PC using oxy-fuel combustion capture and PC using post-combustion capture were approximately at the same cost levels. Based on these results, Kanniche *et al.* recommended taking the following considerations into account during designing near-zero emissions power plants: pre-combustion capture in IGCC, post-combustion in NGCC, and oxy-fuel combustion in PC.

In addition, oxy-fuel combustion is a considerably competitive technique compared to other CO_2capture technologies. Therefore, a large number of experimental and theoretical studies are being carried out in this area worldwide, particularly in developed countries using different scales of furnaces. Generally, these studies depend on solid fuel (coal) because it is a major source of energy in some of these countries. In order to identify any difficulties in using a large-scale boiler during switching from air-fired (conventional) combustion to oxy-fuel combustion, several studies using different scales of furnaces should be widely undertaken. These studies on oxy-fuel-fired scenarios will provide relevant information to maintain the similar combustion characteristics of PC without capture. Thereby, the most cost-effective basis can be achieved for retrofitting existing power plants or to build a new power plant unit under oxy-fuel combustion conditions. However, recent studies have concluded that the oxy-fuel combustion technology is technically feasible and can be applied in a large-scale pulverized coal-fired power plant, and it certainly represents a competitive method relative to the other CO_2 capture technologies. As a result, investigations, developments, innovations and research in this technique are necessary to provide high level confidence and operational experience at a laboratory-scale and pilot-scale and then gradually at a large power plant scale.

3. OXY–FUEL COMBUSTION TECHNOLOGY DEVELOPMENTS

In order to reduce GHG emissions to the atmosphere, systematic research and development work on the existing electricity power plants for CO_2 capture are required. Research into CO_2 capture started two decades ago with both experimental investigations and numerical simulation methods. Lab-scale furnaces were initially used in order for researchers to identify and characterise the fundamentals and operations of oxy-fuel combustion issues under different operating conditions. The fundamental aspects of concern included, for example, flame stability [14, 15], ignition behaviours [16, 17], species concentrations [18], and fuel combustion rate [19]. The heat transfer characteristics under different oxy-fuel-fired scenarios have been studied [20, 21] in order to reduce the retrofits needed to convert conventional boiler designs to oxy-combustion processes. Globally, there have been some studies on pilot-scale oxy-fuel combustion facilities in Europe and some developed countries [22-24]. These investigations revealed that power plants can simply switch from air-fired combustion to the oxy-firing at a large scale and produce higher concentrations of CO_2 in the flue gas. In addition, a significant reduction in NO_x emissions can be achieved due to eliminating N_2 from the oxy-fuel combustion processes [25]. These confirmations and support for oxy-fuel combustion technique for CO_2 capture show that there are no main crucial barriers in implementing this clean, efficient, and economic technology in industrial large-scale facilities. However, switching to oxy-combustion is completely dependent on public support and government developing policies to address global climate change.

3.1. Experimental and Theoretical Laboratory–Scale Projects

With increasing concerns from the Kyoto Protocol about the CO_2 emissions and global warming, research into oxy-fuel combustion technology has widely increased. To overcome the difficulties in applying the oxy-fuel combustion and make it more appropriate and acceptable in the applicable fields, several experimental studies have been conducted with different combustor sizes such as lab-scale and pilot-scale furnaces as the basis for building a large-scale oxy-fuel furnace. In the literature on oxy-fuel combustion, studies on lab-scale furnaces have mainly focused on some important points: ignition behaviour, chemical species characteristics, char combustion, flame propagation speed, NO_x reduction and SO_x formation, and heat transfer models were investigated.

Liu *et al.* (2005) [19] used a 20 kW down-fired coal combustor (190 mm inner diameter and 3 m height) to test the UK bituminous coal combustion in air and in the mixture of O_2/CO_2. The authors showed that the char burnouts and gas temperatures obviously decreased in coal-O_2/CO_2 combustion due to the high specific heat capacity of carbon dioxide compared to nitrogen. They recommended that the concentration of oxygen in O_2/CO_2 mixture should be increased to 30% and 70% for CO_2 (or recycled flue gas) to achieve a corresponding temperature similar to the coal-air combustion.

The influence of reactions, including the effects of char structure and heat transfer, on the ignition behaviour of low-rank Victorian brown coal and high-rank Chinese bituminous coal in air-fired and oxy-fuel combustion cases was experimentally investigated in a wire-mesh reactor by Qiao *et al.* (2010) [26]. As the gas mixture was set at 21% O_2 and 79% CO_2, a slight increase in the average ignition temperature was noted for both the coal types. In contrast, a noticeable decrease in the particle ignition temperature was observed with increasing O_2 concentrations for brown coal and bituminous coal during oxy-fuel-fired scenario. Qiao *et al.* concluded that the reason for that ignition behaviour was the thermal physical properties of the gases surrounding the particles (Qiao et al. 2010). As a continuation of the Victorian brown coal investigation into O_2/N_2 and O_2/CO_2 mixtures, Zhang and co-researchers (2010) [17] and (2010) [27] used a lab-scale drop-tube furnace (DTF). The measurements were conducted using a high-speed camera (MotionPro Y-3) and two-colour pyrometers, the first for photographic observation, and the second for particle temperature measurements. The authors concluded that the coal pyrolysis, coal combustion (volatiles ignition and char oxidation rate), and the surface temperature were highly influenced by the bulk gases in the O_2/CO_2 mixture. There was a clear delay in the coal ignition in the oxy-fuel combustion environment compared to that in the coal-air combustion. This was because a thick volatile cloud released remained attached to the char surface for a long time in the O_2/CO_2 mixture and led, as a result, to the high oxygen consumption on the char surface. Zhang *et al.* recommended increasing the O_2 concentration in the O_2/CO_2 mixture to 27% and 73% for CO_2. This increase leads to achieving good stability in the volatile flame, and obtaining a corresponding char particle temperature to that of the air-fired case.

Computational fluid dynamics (CFD) modelling studies can comprehensively provide a wide range of information for the design of furnace and burner that can reduce the cost of time-consuming experimental investigations. The CFD has the ability to predict well the flame structure, gas temperatures distributions, chemical species concentrations, radiative heat transfer etc., under different combustion conditions. One of the CFD benefits is that the multiple chemistry mechanisms discovered can be used in simulations of the fuel reaction, which are often conducted on the assumption of a chemical balance, finite-rate chemistry scheme, or an approach of mixed-is-burned model [28, 29].

Recently, Venuturumilli and Chen (2009) [30] performed a CFD analysis between the four-step reduced mechanism and its starting mechanism on axisymmetric laminar diffusion (non-premixed) methane flames. The reduced mechanism included 8 major species (CH_4, H, H_2, H_2O, CO, CO_2, O_2, and N_2), while the starting mechanism had 18 species and 65 elementary reactions. The comparison between these two different chemical mechanisms showed that the temperature distributions and axial velocity profiles were similar at the flame base location, whereas the four-step reaction mechanism was not able to provide precise information about the ignition characteristics of the methane flame. Furthermore, the authors estimated the computational time required for the starting mechanism was around 3-4 times as long as for the four-step reaction mechanism. In addition to the above experimental investigations and numerical modelling, the open literature includes a number of relevant findings [31-33] from small lab-scale oxy-fuel furnace experiments. These lab-scale studies under oxy-fuel combustion conditions are useful, and can provide some technical insights and fundamental engineering techniques for this challenging technology. The uncertainties about the heat transfer characteristics, flame behaviour, corrosion problems and pollutant control units in lab-scale furnaces lead to the need for further research on the application of this technology prior to industrial full-scale boiler development.

3.2. Industrial Large–Scale Demonstration Developments

Developments in the lab-scale oxy-fuel facilities as a result of experimental investigations and numerical simulation methods have led to new approaches being used in large-scale facility units. These developments have allowed, effectively, a good compromise between the expensive experimental tests and complete simulation of commercial large-scale with CCS plants. Therefore, the knowledge and technical experience gained from these studies and investigations can be utilized directly to design industrial demonstration large-scale boilers (>250 MW) for the next few years. However, the pollutant control units are not completely understood in oxy-fuel combustion systems, and as a result more research work is required, particularly in the mercury removal units. A lot of information in regard to the international demonstration of oxy-fuel combustion can be found in the International Energy Agency (IEA) Greenhouse Gas Programme, which has been supported by the Asia partnership project. Recently, there have been a number of projects investigating PC oxy-fuel combustion demonstrations, some focussing on electricity production and other on CCS processes in industrial large-scale boilers such as Vattenfall, Endesa, FutureGen 2.0, and KOSEP/KEPRI, as summarized in Table 1 [34]. The implementation of these industrial-scale oxy-fuel combustion projects will greatly reduce the technology costs, especially in the ASU and CO_2 compression equipment, and make it more appropriate for commercial applications in 2022, as reported in the sequence project developments by Wall *et al.* (2009) [1].

Up to date, in the field of numerical simulation on the commercial-scale facility unit, there has been unfortunately little research work conducted on oxy-fuel combustion conditions. Zhou and Moyeda (2010) [35] conducted a process analysis and main calculations on an 820 MW facility to compare the furnace temperature profiles for the air-fired and the oxy-fuel combustion with both wet and dry recycled flue gas modes. The authors proposed several design criteria for both oxy-fuel combustion cases as follows: they would use the same total heat fuel, the same boiler exit O_2 and the same gas flow rate in order to reduce the retrofit impact on the conventional boiler performance. They showed that the flue gas recycle ratio depends on the stoichiometric (air-to-fuel) ratio and ash content in coal. Their results also indicated that the moisture content in O_2/CO_2 combustion was 3.5 times higher than that of the conventional combustion. As a result, a clear reduction in the adiabatic flame temperature was noticed due to the increased moisture content, and therefore 75% of the recycled dry flue gas should be used to increase the flame temperature. The significant resulting increase in the H_2O and CO_2concentrations in the flue gas of O_2/CO_2 combustion cases was accompanied by a clear decrease in NO_x formation due to the reduction in the flame temperature and thermal NO production.

Table 1. Developments of oxy-fuel combustion demonstration projects around the world, adopted from [34]

Project	Location	MW_{th}	Start Up Year	Boiler Type	Main Fuel
B & W	USA	30	2007	Pilot PC	Bit, Sub B, Lig.
Jupiter	USA	20	2007	Industrial	NG, Coal
Oxy-coal UK	UK	40	2009	Pilot PC	Bituminous
Alstom	USA	15	2009	Pilot PC	Bit, Sub B, Lig.
Vattenfall	Germany	30	2008	Pilot PC	Lignite (Bit)
Total, lacq	France	30	2009	Industrial boiler	NG, Coal
Callide	Australia	90	2011	PC-with Electricity	Bituminous (Sub B)
CIUDEN-PC	Spain	20	2010	Pilot PC	Anthra, Bit, Lig.
CIUDEN-CFB	Spain	30	2011	Pilot CFB	Anthra, Bit, Lig.
ENEL HP Oxy	Italy	48	2012	Pilot-High pressure	Coal
Vattenfall	Germany	1000	2014	PC-with Electricity	Lignite (Bit)
Endesa	Spain	1000	2015	CFB-with Electricity	Anthra, Bit, Lig.
FutureGen 2.0	USA	600	2017	PC-with Electricity	Coal
KOSEP/KEPRI	Korea	400	2018	PC-with Electricity	Coal

4. FUNDAMENTALS AND OPERATIONS OF OXY–FUEL COMBUSTION POWER PLANTS

In this section, some of the fundamental issues affecting the operations of oxy-fuel power plant, e.g. coal ignition, flame stability, and char combustion are briefly surveyed. For better understanding of these aspects of the combustion processes, the survey includes both the experimental results and theoretical methods. Other important issues that will be reviewed include the effects some parameters, such as oxygen content, particle size, RFG ratios, and air leakages have on the combustion characteristics. These parameters are considered to have a significant influence on the performance and reliability of oxy-fuel power plants.

4.1. Coal Ignition

To understand the impact of the specific heat capacity of the gas mixture on the coal ignition process, the following mathematical equations can be inferred from the above observations. Once the volatile matter has been released during the volatilization process, the auto ignition time of coal particles can be determined based on the assumptions of ignition and explosion theory [36]. For a one-step overall reaction and no heat loss, the ignition delay of volatile gases is given as follows:

$$\tau = \frac{c_v}{Q_c Y_{F,O} A \exp(-T_a / T_o)} \frac{T_o^2}{T_a} \quad (11)$$

Where, cvcv is a specific heat capacity at constant volume, QcQc is the combustion heat release per mass of fuel, YF, OYF, O is the mass fraction of fuel at initial time value t=0, ToTo is the initial temperature of reactants, TaTa is the ambient temperature and AA is a kinetic factor in the Arrhenius expression. According to Eq. 1, it is clear that the ignition delay time increases with an increase in the specific heat capacity (cvcv) of gases and decreases in proportion to the combustion heat release (QcQc). Therefore, the high specific heat capacity can play an important role in increasing the ignition delay. As mentioned earlier, the specific heat capacities of the gaseous combustion products such as CO_2 and H_2O are higher than that of N_2. As reported in many oxy-fuel experimental investigations [16, 37, 38], the ignition delay can be reduced by increasing the oxygen concentration in the gas mixture in order to enhance the chemical reaction rate, as explained in Eq. 1. As a result, a higher O_2 concentration can be used to dilute the carbon dioxide effect on the ignition mechanism and to yield the same ignition time for oxy-fuel combustion as that of coal-firing in the air.

In addition, the chemical effects on the coal particles, due to the elevated concentrations of CO_2 and H_2O in the flue gas of oxy-fuel combustion, have been considered as another reason for the coal ignition delay. This chemical phenomenon was observed in two experimental studies, which were conducted in lab-scale drop-tube furnaces (DTF) in Australia [39, 17]. The authors used different types of Australian coals and a wide range of O_2 levels to investigate the coal pyrolysis behaviour, ignition extent, and char burnout in air-fired and oxy-fuel-fired environments. Zhang *et al.* (2010) [17] noted that when the nitrogen is replaced by carbon dioxide in the gas mixture, this enhances coal pyrolysis prior to ignition and as a result produces a large cloud of thick volatile matter on the char surface. Thereby, ignition of the volatile cloud occurred instead of single particle ignition. Approximately 30% of O_2 concentration in the O_2/CO_2 mixture was recommended to be used in order to achieve similar coal ignition like that in the O_2/N_2 mixture. Rathnam *et al.* showed that the apparent volatile yield measured in the DTF for O_2/CO_2 mixture at 1673 k was around 10%, as high as that in air-firing for all coal types.

4.2. Flame Behaviour

In general, the physical parameters of the flame, for instance shape, brightness, extinction, oscillation frequency, and temperature level, can be used to characterise the flame behaviour and control the stability of flame in the furnace to improve combustion efficiency. As a consequence, flame stability is an important issue and needs to be studied and taken into consideration when designing burners in order to improve flame characteristics and reduce emission levels. The replacement of air by CO_2 in the feed oxidizer gases has been found to have a significant effect on flame stability. However, to maintain a better flame stability in an oxy-fuel furnace, Chui *et al.* (2003) [22] tested two different burners' configurations (A and B) to compare which has optimal efficiency. Both experimental investigation and numerical modelling studies on a pilot-scale furnace were conducted. The main difference between the designs of the two burners was the location of the pure oxygen injection into the primary gas stream. For burner A, the annulus high-O_2 jet was located inside the primary stream, while in burner B, the annulus high-O_2 jet was located between the primary and secondary streams and without a cyclone chamber for coal delivery. The experimental and numerical results showed that burner A improved the flame stability and achieved a significant decrease in the NO_x level in the combustion flue gas in comparison with burner B, particularly when the swirl number was increased in the secondary stream. The improvement was due to the enhancement of the internal recirculation zone of reactants in the near-burner region.

Using state-of-the-art turbulence-chemistry interaction, several advanced turbulent combustion models have been developed such as Eddy-Breakup (EBU) model, Probability Density Function (PDF) transport model, and Conditional Moment Closure (CMC). Most of these mathematical combustion models have only been utilized for the calculations of air-fired flame. For that reason, Kim *et al.* (2009) [40] adopted the CMC model to analyse the characteristics of turbulent combustion of natural gas (NG) flame in air-fired and oxy-fuel combustion environments. The authors coupled the CMC model with a flow solver. The detailed chemical kinetics mechanism model was implemented to calculate the intermediate species (CO and H_2) formed in the flame envelope because of the enhanced thermal dissociation. The normalized enthalpy loss variable (ξξ) used to calculate the effect of convective and radiative cooling terms on the turbulent flame, and it is defined as follows:

$$\xi = h - h_{min} / h_{ad} - h_{min} \tag{2}$$

Where, hadhad is the conditional adiabatic enthalpy, and hminhmin is the conditional minimum that is calculated when the conditional temperatures are reduced to the surrounding temperature.

The numerical results showed that the oxy-fuel flame produced a much broader region in the hot-flame zone, particularly in the lean-fuel side of the mixture fraction, in comparison with the air-NG flame. They also noted that the temperature value of oxy-fuel flame is higher than that of air-firing. The predicted temperature levels for both combustion cases were overestimated compared to the experiments. That discrepancy arose because of the inadequacy of the modified K–εK-ε turbulence model used in the calculations and to the measurements uncertainties. In order to mathematically demonstrate the chemical effect of CO_2 on the reduction of propagation flame speed, the extinction theory of diffusion and non-diffusion flames will be considered. The Damkohler number (D) is a dimensionless number, and can be used to interpret the extinction characteristics of flames, as follows [41, 42]:

$$D = \frac{\tau_r}{\tau_{ch}} \tag{3}$$

Where τrτr represents the residence timescale of reactants in the combustion zone, also known as the mixing timescale, and τchτch is the chemical reaction timescale and it is equal to a ratio of the characteristic thermal diffusivity of the gas divided by the square laminar burning speed (i.e. $\tau_{ch} = \alpha / S_L^2$). The numerator of the D number is completely dependent on the fluid dynamics of the flame, but the denominator of D is a function of the flame reaction rate. Therefore, the mathematical expression of the Damkohler number can be rewritten using the Arrhenius kinetic rate formula, as follows:

$$D \propto \frac{\tau_r}{\alpha} exp(-E_A / R\, T_f) \tag{4}$$

According to Eqs. 3 and 4, flame velocities in oxy-fuel combustion environments will observably be lower than those in air-fired environment at the same level of oxygen concentration. This is due to the lower values of thermal diffusivity and adiabatic flame temperature in O_2/CO_2 mixtures compared to those in O_2/N_2 mixture. As a result, the flame stability of the oxy-fuel combustion will definitely be affected by that reduction attained in propagation flame speed.

4.3. Char Combustion

After the devolatilization process of the coal particles finishes, the char combustion subsequently starts in the firing system. This combustion process has been considered the dominant factor constraining several reaction parameters such as the total burnout time, unburned carbon level, and radiation from burning char particles. A better understanding of the effects of these parameters on the combustion characteristics and boiler heat transfer under oxy-fuel conditions enables engineers optimize the applications of both the existing and new coal power plants.

In order to reduce the char burnout time and increase the combustion rate of char particles in O_2/CO_2mixtures, oxygen-enriched atmosphere must be used in the gas mixture [23]. However, the literature shows that the combustion rate affects the reaction order of the bulk oxygen partial pressure. This means that the combustion rate when measured in terms of the chemical kinetic control reaches a high reaction order (0.6-1) for O_2 below 900K, but a low reaction order in the oxygen partial pressure is exhibited when the global reaction rate is above 1200K [43]. Based on these findings, the reaction rate of coal char particles has been interpreted to be subject to an n-th order law governing oxygen partial pressure and the Arrhenius kinetic rate model of char reaction. The overall rate of gasification is given as follows:

$$r_{gas} = k_s(T_p)\, P^{n}_{O_{2,s}} \tag{5}$$

Where nn represents the reaction order, TpTp is the temperature of the coal particle, and ksks is a coefficient of temperature dependent rate and can be written according to the Arrhenius expression: $k_s(T_p)=A\, exp(-E\,/\,R\,T_p)$.

The elevated CO_2 and H_2O concentrations in the flue gas of oxy-fuel combustion will influence the reaction rate of char particles. As previously mentioned, this can be attributed to the relatively high specific heat capacities of the main gaseous products (CO_2 and H_2O) compared to the nitrogen in the air-firing. These dominant species will act with the remaining char particles following the endothermic process and thereby reduce the char particle temperature, resulting in a decrease in the rate of char oxidation. Furthermore, the diffusivity of oxygen on the char surface in an O_2/CO_2 mixture is lower than that in an O_2/N_2 atmosphere, and this could be why a lower char oxidation rate is observed in an O_2/CO_2 mixture. Due to the importance of the elevated gases on the gas temperature and char burning rate, Hecht *et al.* (2010) [44] used a computer model to study the effect of the endothermic CO_2gasification reaction on the char consumption under different oxygen-enriched environments. Numerical modelling was implemented employing the Surface Kinetics in Porous Particles (SKIPPY) code over a range of potential CO_2 oxidation rates for bituminous coal particles. The SKIPPY depends on the FORTRAN program to solve the conservation mass, momentum, energy, and species concentration equations by assuming a multicomponent gaseous phase. The results showed that where there was 12% oxygen in an O_2/CO_2 mixture, the endothermicity of the CO_2 gasification led to a significant decrease in the char particle temperature, and thus reduced the reaction rate of char oxidation. For up to 24% O_2, the global consumption rate of char particles enhanced with increasing the reaction rate of CO_2 gasification. With more than 24% O_2, the overall rate of char reaction decreased with an increase in the gasification rate of carbon dioxide.

4.4. Effect Of Some Parameters On Oxy–Fuel Characteristics

When a mixture of pure oxygen and recycled flue gas is used instead of air as the combustion gases in oxy-fuel-fired power plants, many modifications to the combustion characteristics of the power plant boiler will occur under normal operating conditions. Therefore, in order to address these challenges and ensure the oxy-fuel power plants are working at high combustion efficiency, consistent thermal performance, and with low emissions, several major parameters can be utilized to achieve that remarkable goal. Based on the design and operating conditions of the existing power plants, the effects of different parameters under oxy-fuel combustion conditions such as oxygen concentration, particle size, and recirculation of flue gas have been investigated. In this subsection, a brief summary of research under the effects of these different parameters is presented.

4.4.1. O_2 Concentration

The oxygen concentrations in the mixed oxidizer stream have a significant impact on the flame stability and heat transfer characteristics in the oxy-fuel firing facilities. In order to maintain the same aforementioned characteristics as in the conventional firing systems, the desirable value of the O_2concentration has to be precisely determined. Due to the physical and chemical differences between the properties of carbon dioxide and nitrogen, it seems that 21% O_2 concentration (by volume), in a mixed stream of oxy-fuel conditions, does not provide the same combustion characteristics as conventional operations. This can be explained, as mentioned earlier, as a result of the delay in the ignition time of volatile matter released and to the difficulties in oxidation of the coal char particles. Therefore, higher oxygen concentrations are required to work safety and provide more efficient operations. However, the RFG enriched with 28 vol. % O_2 is safer than of 21% O_2 in the secondary stream of a pilot-scale burner in oxy-fuel combustion tests for high-volatile subbituminous coal [22].

The oxy-fuel combustion scenarios offer the opportunity to supply different amounts and concentrations of RFG enriched with O_2 to both the fuel carrier gas and feed oxidizer gas streams. Many studies have shown that a range between 25% and 36% oxygen (by volume), at the burner inlet, is preferable in oxy-fuel-fired conditions to maintain the same flame behaviour and heat transfer characteristics as those in the air-fired [14, 17, 45-47]. In addition to the safety reasons, this range of oxygen content in the gas mixture was basically found to result in a lower level of pollutants in the flue gas such as NO_x, SO_x, CO, and trace elements during the burning of several coals types. By increasing the O_2 concentrations to more than 21 vol. % in both the primary transport gas and RFG streams at the commercial low NO_x swirl burner, Sturgeon *et al.* (2009) [33] noted some improvements were achieved in different areas of investigation compared to the standard conventional firing. For example, the level of CO concentration was decreased, a stable flame was attained, and a low content of carbon in the ash was noticed.

4.4.2. Particle Size

The findings on the effect of coal particle diameters on the oxy-coal firing have shown that there are recognizable influences on the flame propagation speed, devolatilization process, and ignition temperature. Suda *et al*. (2007) [47] used two different particle diameters (50 μm and 100 μm) to investigate the effect of coal particle size on the propagation behaviour of flames. The results showed that the flame propagation velocity slightly decreased with coal particle size. However, the reason for that decrease is the lower heat transfer conduction process between the coal particles and gas. Moreover, the authors concluded that the flame stability in the O_2/CO_2 mixture could be increased using smaller PC particles.

Huang and co-researchers (2008) [46] investigated the effect of the coal particle size on the combustion characteristics of TieFa (Tf) bituminous coal in different mixtures of O_2/CO_2 atmospheres during the experimental tests. The experiments were conducted using a differential thermal analyser (DTA). Three different particle sizes (11.34, 18.95, and 33.68 μm) and four different O_2concentrations (10, 20, 50, and 80%) in the gas mixture were used in order to carry out comprehensive investigations into the effect of these two important parameters. The authors used a ratio of thermal gravimetric (TG) to the differential thermal gravimetric (DTG) to clarify the combined effect of the above-mentioned factors on both the devolatilization and char combustion processes. The results of the TG/DTG curves showed that the weight loss of coal samples was augmented as the coal particle size decreased. That contributed to an increase in the surface area of coal particles that led, as a result, to enhance the overall reactivity of coal char, particularly when the O_2 concentration was increased from 10% to 80% at a constant heating rate. In addition, there was a clear decrease in both the ignition and burnout temperatures at the small coal particle size. The effect of the particle size on the reaction rate was relatively negligible at oxygen-rich conditions (50% and 80%). The combustion property index (SS) was also used to give an inclusive estimation of the combustion characteristics under these specified conditions and practical operations. The index (SS) was defined as the coal reaction rates divided by the square of ignition temperature and burnout temperature, as follows [48]:

$$S = \frac{\left(\frac{dW}{dt}\right)_{max}^{c} \left(\frac{dW}{dt}\right)_{mean}^{c}}{T_i^2 T_b} \tag{6}$$

Where (dWdt)cmaxdWdtmaxc and (dWdt)cmeandWdtmeanc represent the coal burning rates at the maximum and mean values, respectively. The SS values were determined with a constant heating rate of 30 ℃°C/min. The plotting of the combustion property index against the coal particle sizes show that the index SS not only increased with a decrease in the particle size, but also with an increase in the oxygen concentration. Consequently, they concluded that the SS values become superior with these two intensified combustion parameters [46].

4.4.3. Flue Gas Recirculation

The recirculation of flue gas or recycled flue gas (RFG) is an important process in oxy-fuel combustion. In this process, a large amount of combustion flue gases is recycled to the furnace in order to maintain the same combustion temperature and heat transfer characteristics in the boiler as in the conventional coal-fired power plant. The recycle ratio of flue gas can be defined according to the following simple mathematical expression [49]:

$$Recycle\ ratio = \left(\frac{Recycled\ gas\ mass\ flow\ rate}{Recycled\ gas\ mass\ flow\ rate + Product\ gas\ mass\ flow\ rate}\right) * 100 \tag{7}$$

Typically, 60% to 80% of produced flue gas, mainly CO_2 and H_2O, is recycled in the oxy-fuel plants, and this ratio is basically dependent on the coal type and the options of RFG [35]. As previously illustrated in Fig. 4, the oxy-fuel technique offers two options to draw the flue gas, either wet RFG or dry RFG can be practically used for the same purpose, depending on from which locations the flue gas is taken from the system. Generally, when using the wet RFG option, the produced combustion gas is extracted before the condensing process, whereas the dry RFG is extracted downstream from the condenser, and both of these options are carried out after the removal of particulates. Under the same volumetric flue gas flow rate, the dry RFG has higher adiabatic flame temperature (AFT) than that in the wet RFG. However, the combination of wet and dry RFG options may provide an attractive option, especially for the PC oxy-fuel power plants. The dry RFG can be utilized to transport coal particles from the mill to the furnace and other miscellaneous uses, while the wet RFG is used to control the combustion temperature due to its high content of water vapour which is much more than in dry RFG. This combined attractive option has recently been used by the Vattenfall project in Germany on the 30 MW oxy-lignite pilot-scale utility boiler [50].

Sturgeon *et al.* (2009) [33] demonstrated the effect of the RFG ratio on the amount of furnace exit carbon in ash (CIA), on the adiabatic flame temperature, and on the coal residence time in the combustion zone. By increasing the RFG ratio, the flame temperatures and residence times of PC were gradually decreased, resulting in an increase in the exit of CIA from the furnace. In order to avoid making major modifications to the heat exchange equipment of the conventional large-scale boiler, the recycle ratio of flue gas, in the oxy-fuel scenario, has to be precisely determined. The best value of RFG can bring another benefit to the oxy-fuel power plant through eliminating the slagging and fouling formation problems on the water wall and on the convective tubes bank of the boiler such as superheaters and reheaters. This problem can be avoided by keeping the furnace exit gas temperature below the ash melting temperature of coal combustion [10, 12, 16]. This specific temperature is substantially dependent on the ash depositing behaviour of the coal used, as will be seen in the discussion sections of chapters five and seven.

5. CFD MODELLING INVESTIGATIONS ON THE LAB–SCALE AND LARGE–SCALE FURNACES UNDER AIR–FIRED AND OXY–FUEL COMBUSTION CONDITIONS

In this section, an overview of the research program [10, 11, 51, 52] will be briefly discussed. This research program can be classified to have two main objectives:

- The 3 D numerical simulations of pulverized dry lignite in a 100 kW test facility were conducted to simulate four different combustion environments (air-fired, OF25, OF27, and OF29) and to investigate the temperature distribution levels, species concentrations, and velocity. The commercial CFD software was used to model and analyze all the combustion media. Several mathematical models with the appropriate related constants and parameters were employed for lignite coal combustion. The combustion conditions of oxy-fuel combustion cases were satisfied by modifying the following factors: oxygen concentrations in the feed gas and carrier gas, and recycled flue gas rates [10, 51].

- The commercial CFD code was modified to investigate the Victorian brown coal combustion in a 550 MW tangentially-fired boiler under different combustion media. Several mathematical models such as coal devolatilization, char burnout, combustion chemistry, convection and radiation heat transfer processes, carbon in fly-ash, and thermal and fuel nitric oxides models were developed through subroutines and added to the CFD calculations. The available experimental data from the power plant were used to validate the predicted results under air-fired condition; a good agreement was achieved. The oxy-fuel combustion approach adopted in a 100 kW facility unit (Chalmers' furnace) was applied to the present large-scale furnace in three O_2/CO_2 mixture conditions, namely OF25, OF27, and OF29. These models were implemented to investigate the importance of including such models in conjunction with the newly developed oxy-fuel combustion model [52].

5.1. The Pulverized Dry Lignite Combustion In The Lab–Scale Furnace

This subsection describes a comprehensive computational fluid dynamics (CFD) modelling study undertaken [10, 51] by integrating the combustion of pulverized dry lignite in several combustion environments. Four different cases were investigated: an air-fired case and three different oxy-fuel combustion environments (25 % vol. O_2 concentration (OF25), 27 % vol. O_2 concentration (OF27), and 29 % vol. O_2 concentration (OF29). The chemical reactions (devolatilization and char burnout), convective and radiative heat transfer, fluid and particle flow fields (homogenous and heterogenous processes), and turbulent models were employed in 3-D hybrid unstructured grid CFD simulations. The available experimental results [53] from a lab-scale 100 kW firing lignite unit (Chalmer's furnace) were selected for the validation of these simulations.

5.1.1. A Review Of Experimental Setup Of Andersson [53]

Chalmers 100 kW test facility has been designed to burn both gaseous and pulverized fuels. The furnace is a cylindrical refractory lined drop tube unit with dimensions: 80 (cm) inner diameter and 240 (cm) inner height. Three different oxy-fuel combustion cases are used as follows: OF25 (25% vol O_2 and 72% vol CO_2), OF27 (27% vol O_2 and 71% vol CO_2), and OF29 (29% vol O_2 and 69% vol CO_2). The carrier gas in the air-fired case, which is used to inject the coal, is air with a volumetric flow rate of 83.3*10^{-5} (m^3 per sec). While the carrier gas in oxy-fuel combustion cases is recycled flue gas (RGF) with a volumetric flow rate of 66.6*10^{-5} (m^3 per sec.) that has the same oxygen concentration (30% vol dry) and recycled flue gas (68% vol dry) for all oxy-fuel combustion cases.

Table 2. The inlet flow field parameters of all combustion cases for primary and secondary registers of the burner

Inlet Flow Field Parameters		Combustion Cases			
		Air	OF25	OF27	OF29
Primary Register	Volume Flow Rate (m^3/h)	34.87	28.94	26.85	25.11
	Mean Velocity (m/s)	7.966	6.612	6.134	5.737
	Angular Velocity (rad/s)	433.293	359.615	333.645	312.052
Secondary Register	Volume Flow Rate (m^3/h)	81.37	67.54	62.65	58.59
	Mean Velocity (m/s)	4.995	4.146	3.845	3.596
	Angular Velocity (rad/s)	41.14	34.152	31.673	29.608

Generally, the volumetric flow rates of feed oxidizer gases (air or dry RFG for oxy-fuel combustion cases) were 30% for primary register and the rest were through secondary register. The volumetric flow rates of air or RFG through the primary and secondary registers of the burner were decreased gradually by 17%, 23%, and 28% in the OF25, OF27, and OF29, respectively with respect to the air-fired case. The oxidizers / fuel stoichiometric ratio (λ) was kept constant (1.18) for all combustion cases. The initial inlet temperature of dry flue gas at the inlet of burner was around 298.15 K for all combustion tests. The inlet flow field parameters at the primary and secondary registers of the burner, for all the combustion cases, are summarized in Table 2. The combustor is initially fired by using gas (propane) as a pilot-fuel to start up the combusting process before switching to coal.

5.1.2. Cfd Results and Discussion

Figure 4 shows the temperature distributions on a vertical plane through the middle of the furnace. The left hand sides of each image show the air-fired case as the reference. The three different oxy-fuel combustion cases are presented on the right hand sides. The main purpose of this figure is to visualize the overall flame temperature in this axisymmetric furnace.

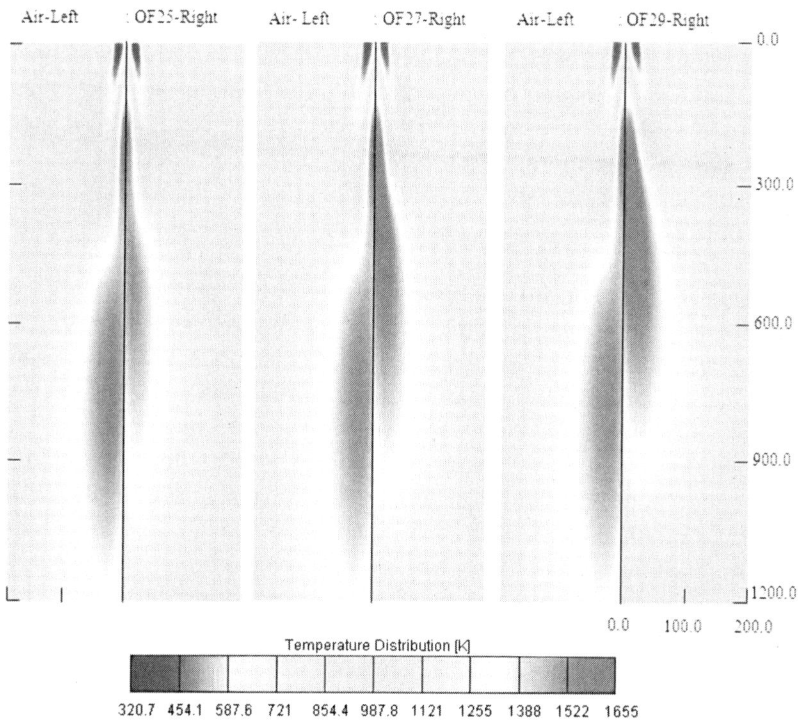

Figure 4. Flame temperature distributions (K) at the vertical cut along the furnace axis for the reference air-fired case (left hand sides) and oxy-fuel combustion environments OF25, OF27, and OF29 (right hand sides), respectively.

The flame started at the burner exit and extended up to the middle of the furnace for the reference air-fired case. The near-burner flame temperature distribution of oxy-fuel combustion (OF25) is closer to that of the air-fired case. Whereas, the near-burner flame temperature of oxy-fuel (OF27) is higher, and the length of the flame is shorter and more confined in the burner exit region. For the OF29 case, the flame temperature distribution is similar to that of OF27 case with marginally higher flame diameter in the near burner region. The peak flame temperature values of the air-fired and OF25 cases were 1603.3 (K), and 1577.5 (K) respectively. While the peak flame temperature values for OF27 and OF29 were 1666.2 (K) and 1699.1 (K), respectively. These peak flame temperature values are very consistent with the experimental result values [53]. It is clear that when the oxygen concentration in the feed gas stream is increased and the recycled flue gas (RFG) is decreased in the oxy-fuel combustion environments, the flame temperature increases and the shape of the flame is more confined and close to the inlet flow field tip. This phenomenon can be attributed to the higher oxygen content in the near burner reaction zone and higher residence time for the coal due to lower velocity that gives more time to burn. Furthermore, the decreased amount of recycled flue gas (CO_2) will absorb less heat released by combustion, thereby increasing the flame temperature.

In Figure 5, the velocity vectors at the primary and secondary swirl registers, located at the tip of the burner. The differences in the directions of velocity vectors between the primary and secondary registers are related to the values of swirl numbers at the same combustion case. The values of the velocity vectors for both inlet registers are different for different combustion cases and are dependent upon the inlet flow conditions as reported in Table 2. A swirl injection system is widely used in the burning systems in order to increase the mixing and give enough time for oxidizers to burn maximum amount of fuel in that critical zone of the furnace thereby avoid incomplete combustion. The swirl effect, in this study, is certainly used to enhance the turbulent mixing and thereby leads to stabilize the structure of the flame.

The oxygen concentrations (mass fraction, kg/kg) on a vertical plane through the middle of the furnace are shown in Figure 6. In this figure, the left hand sides of the images are for the reference air-fired case, while the right hand sides are for OF25, OF27, and OF29 respectively. The oxygen concentrations in the air-fired case and OF25 case are approximately similar. These two cases show a delay in the consumption of O_2 in the upper part of the furnace, especially along the centerline of the furnace, compared to OF27 and OF29 cases. The similarity in O_2 consumption between air-fired and oxy-fuel cases is approximately coupled to the flame temperature levels described in Figure 4. In both the OF27 and OF29 cases, the O_2 consumption starts early due to improved ignition conditions and faster combustion leading to the flame to be closer to the burner tip. These results of oxygen concentrations are very similar to that obtained in experiments [53].

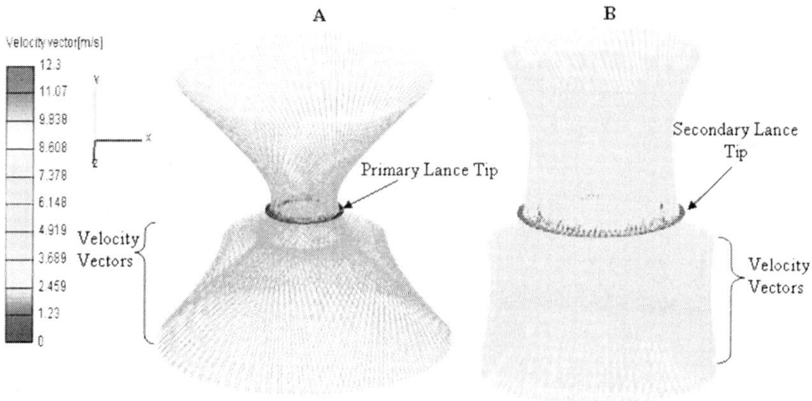

Figure 5. The velocity vectors (m/s) of the primary (A) and secondary (B) swirl registers of the burner in the inlet flow tip of the furnace.

The rapid reduction of oxygen concentration in the near burner region, due to burning of larger amounts of volatile in the OF27 and OF29 cases, negatively affected the oxidation of the residual char in the remaining part of the furnace. Figure 7a shows the char content of coal particles (%) for the air-fired (A) and OF25 (B) combustion cases in the top half of the furnace. Availability of sufficient oxygen led to burn out of residual char by about 900 (mm) into the furnace. Figure 7b demonstrates the residual char for OF27 (C) and OF29 (D) cases. Shortage of the oxygen content led to reduced char burnout, particularly in the OF29 combustion case. The exclusion of the carbon monoxide (CO) as chemical species in the combustion process (i.e. ignoring the Boudouard reaction) of coal, especially in the flame envelop zone (higher temperature region) may have resulted in lower char burnout in the OF27 and OF29 cases. However, these results certainly confirm that the burning out of hydrocarbon gas is faster and low char burnout is achieved at high O_2 concentration of the oxy-fuel combustion cases as reported in the previous findings [17, 19, 24, 54]. As a result, there will be some differences in the total radiation intensities inside the combustor between the reference air-fired case and oxy-fuel combustion case at high O_2 content, especially with high CO_2 concentration in the flue gas.

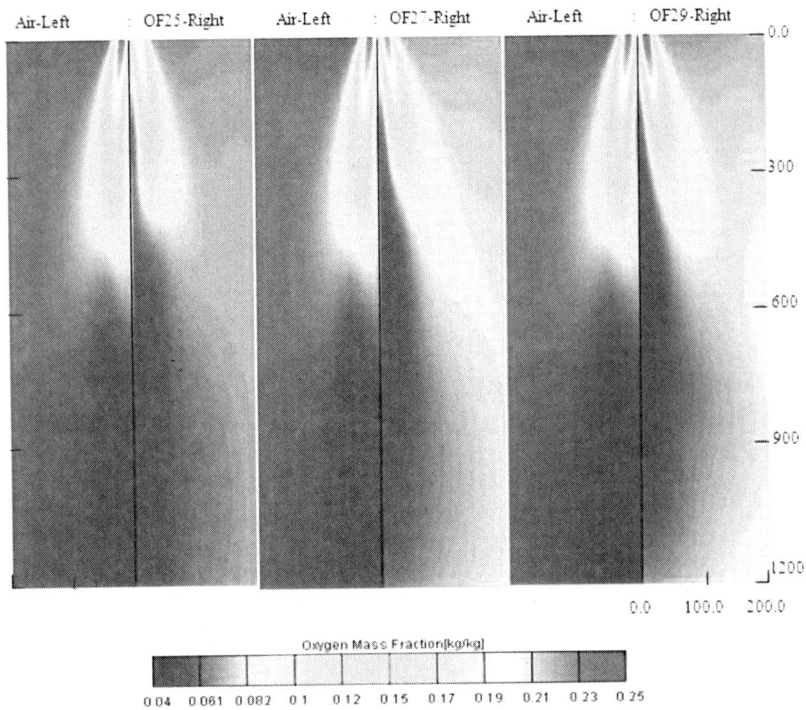

Figure 6. Oxygen mass fraction (kg/kg) in the upper part of the combustor for the reference air-fired case (left hand sides) and oxy-fuel combustion cases OF25, OF27, and OF29 (right hand sides), respectively.

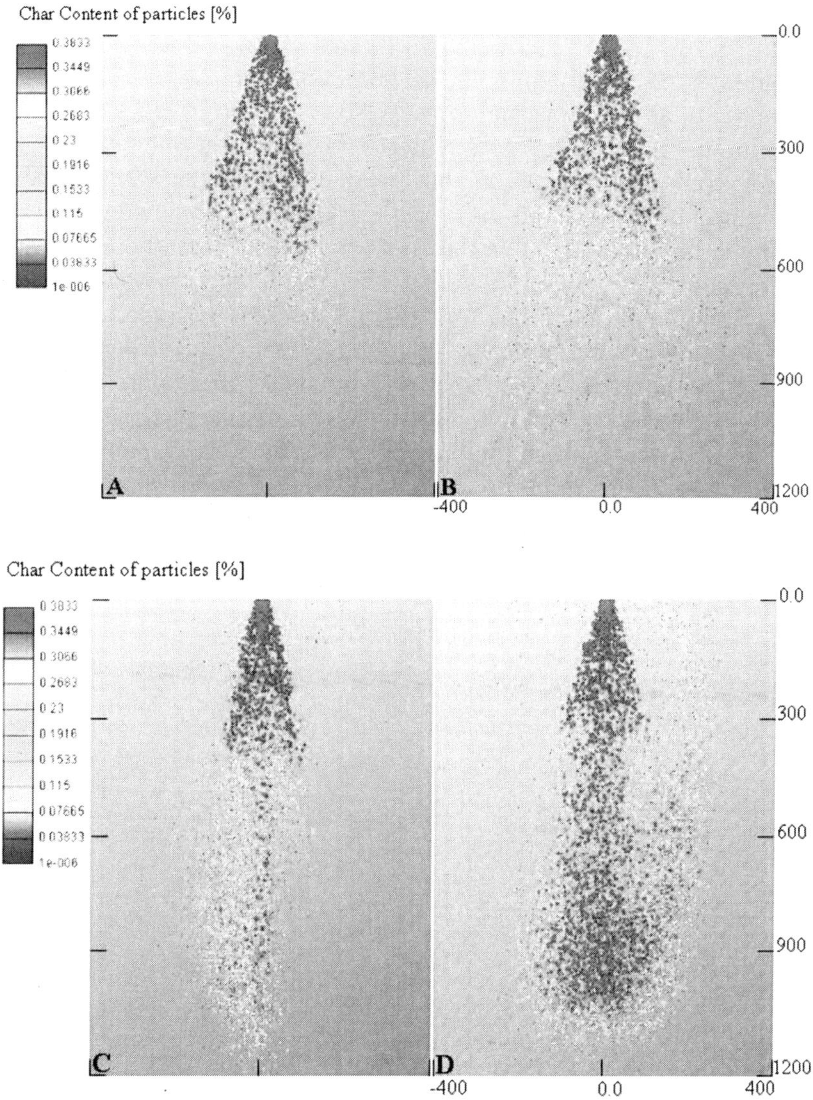

Figure 7. a. Char content of coal particles (%) for the air-fired (A) and OF25 (B) combustion cases over a cross-section of the upper half of the furnace, b. Char content of coal particles (%) for the OF27 (C) and OF29 (D) combustion cases over a cross-section of the upper half of the furnace

The fundamental concept of oxy-fuel combustion technique is mainly to increase carbon dioxide concentration in the flue gas. This technique can be applied, as previously mentioned, by using a mixture of pure oxygen and part of RFG as feed oxidizer gases instead of air to burn with fuel. Therefore, in this study, Figure 8 is clearly showed the increase in CO_2 concentration for all oxy-fuel combustion scenarios (OF25, OF27, and OF29) with respect to the air-fired case. These results were obtained with three-step chemistry mechanisms in terms of homogeneous and heterogeneous coal reactions. As seen, the maximum mass fraction value of CO_2 concentration was about 17.21 % (kg/kg) for air-firing, while for oxy-fuel cases was, in general, about 90.11 % (kg/kg) due to usage of dry flue gas recycled, as implemented in experimental work of Andersson (2007). However, the purity of oxygen (99.5% pure oxygen used in the experiments) and leakage in the furnace are relevant parameters to decrease CO_2 concentration, and therefore they should be taken into consideration in design any oxy-fuel combustion boiler. All these CFD results were comprehensively validated against the available experimental data [10, 51].

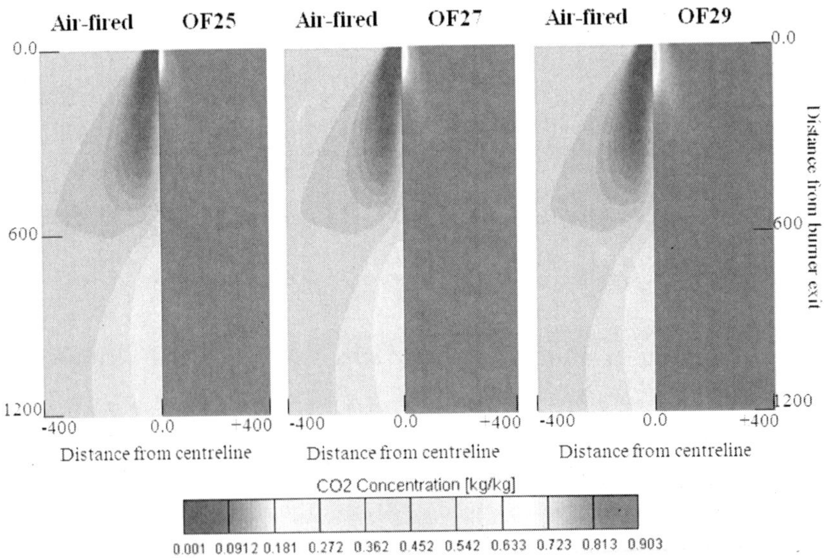

Figure 8. Carbon dioxide concentration (kg/kg) at the vertical cut along the furnace axis for the reference air-fired case (left hand sides) and oxy-fuel combustion environments OF25, OF27, and OF29 (right hand sides) respectively, all dimensions are given in mm.

5.2. The Brown Coal Combustion In A 550 Mw Tangentially–Fired Boiler

In order to design such efficient, clean, and economical brown coal combustion systems, the understanding of the brown coal reactivity and behaviour under several combustion conditions is required. Generally, brown coal has a number of advantages such as abundance, low-cost, high reactivity, and low sulphur content. In despite of these benefits, a high moisture content (about 60-70 wt %) is the major disadvantage of brown coal. However, in the existing pulverised brown coal (PC) tangentially-fired boiler, a large amount of the hot exit flue gas, typically 50% of the total flue gas generated, is reused to dry the brown coal within the mill-duct system [55- 57]. During that drying process by the hot gas off-takes (HGOTs), a large amount of water vapour is reproduced as well. In order to avoid any flame stability problems inside the combustion chamber, due to that evaporated steam, a fuel-rich mixture (mainly pulverised coal) is passed through the main burner ducts. Whilst a fuel-lean mixture, including water vapour, inert gases, and remaining of PC, is delivered to the inert burner ducts (upper burners). This distribution of the PC and inert gases into the firing system is favourable, particularly in this type of combustion technology. This section focuses on the numerical investigation of a large-scale oxy-fuel furnace. Therefore, the objective of this study is to simulate the brown coal combustion in the large-scale tangentially-fired furnace under several operating conditions. A computational fluid dynamics (CFD) code was used to model and analyse four different combustion environments. A conventional PC combustion and three oxy-fuel combustion scenarios, are known as OF25, OF27, and OF29, were simulated. The validation of the CFD results with the power plant data has been conducted in the air-fired combustion. Four parameters: flue gas composition, gas temperatures, carbon in fly-ash, and HGOT mass flow were compared. Results, for all combustion cases investigated, are compared. The species concentrations, temperature distributions, gas-phase velocity fields, char burnout, NO_x emissions, and radiative heat transfer obtained for all combustion cases were compared.

5.2.1. Boiler Description And Operating Conditions

The tangentially-fired Victorian brown coal 550 MW_e boilers located in the Latrobe Valley mine, Victoria/Australia was used in this simulation study. The geometric description of the CFD model for the boiler, unit 1 at Loy

Yang A, is shown in Figure 9. Under maximum continuous rating (MCR) of operating conditions, the unit produces 430 kg/s of steam flow through the main steam piping at 16.8 MPa and 540 °C. The computational domain illustrated in Figure 9 was extended from the furnace hopper up to the top of the tower, passing through the transition of round duct to before the bifurcation at the inlet to the air heaters. In this CFD model, the complex geometric dimensions of the simulated boiler were 98.84 m (height), 17.82 m (width), and 17.82 m (depth), having a net simulated volume of 35,894 m^3 up to the bifurcation point to the air heaters. The tangentially-fired furnace used in this study consists of eight mill-duct systems, two on each side face of the four-sided furnace. For each mill-duct system, there are six separate burners, including three inert burners and three main burners, as well as a hot gas off-take (HGOT) that dries the brown coal. The mill-duct systems were designed for the following purposes: grinding the raw coal into pulverised coal (PC) in the mill, removing the moisture content (62% wt) from the brown coal through the drying shaft, and transporting and distributing the PC. The centrifugal separation system is used to deliver pulverised coal from the grinding mill to the inert and main burners of the furnace. The distribution of PC at both the burner mouths was accompanied by the inert flue gas and water vapour from the drying process in the mill. Approximately 82% of the PC and 34% of the gases is delivered to the main burner (PC burner) and the remaining 18% of pulverised coal and 66% of the gases is transported to the inert burner (vapour burner). This distribution of fuel and gases (fuel-rich mixture) to the main burners is required to maintain combustion stability in the furnace burning Victorian brown coal. Table 3 shows the mass flow distribution of PC and mill gas at each inlet port of the burner ducts. The overall number of vapour and PC burners was 48, while 18 of the total burners were practically out of service, i.e. no fuel is introduced in the latter burners. In the furnace zone, the burners' arrangements on the furnace wall surface were as follows from top to bottom: upper inert burner (UIB), intermediate inert burner (IIB), lower inert burner (LIB), upper main burner (UMB), intermediate main burner (IMB), and lower main burner (LMB).

5.2.2. Cases Studies Set Up

For this numerical study, four different combustion scenarios were selected in order to estimate the performances of the 550 MW_e large-scale boilers under different firing conditions. In the first simulation case, the chemical and physical set up of the boiler operations were completely based on the station data [58]. This first case represents a reference (air-fired) case in investigating the behaviour of brown coal combustion and the boiler performance for the three challenging oxy-fuel combustion scenarios. For the proposed (retrofitted) oxy-fuel combustion scenarios, the thermodynamics set up of the lab-scale oxy-fuel furnace [53], conducted at Chalmers University, was selected in terms of the gas compositions and volumetric flow rates of feed oxidizer gases. The retrofitted oxy-fuel combustion cases were defined as follows: OF25 (25 vol.% O_2 and 75 vol.% CO_2), OF27 (27 vol.% O_2 and 73 vol.% CO_2), and OF29 (29 vol.% O_2 and 71 vol.% CO_2). Detailed information about the Chalmers' furnace and the combustion conditions can also be found in the previous simulation studies [10, 11,51].

Table 3. The mass flow rates and the distribution ratios for PC and mill gas at each inlet port of the burner ducts

Burner duct	PC flow (as receive.)		Gas flow	
	Mass flow rate (kg/s)	Distribution ratio (%)	Mass flow rate (kg/s)	Distribution ratio (%)
UIB	0.93	5.7	18.71	22.1
IIB	0.54	3.3	18.22	21.5
LIB	1.57	9.4	18.19	21.5
UMB	4.82	29.7	10.3	12.2
IMB	2.78	17.2	9.34	11.1
LMB	5.62	34.7	9.8	11.6
Total	16.26		84.56	

5.2.3. Cfd Results and Discussion

In Figure 10, the distributions of flue gas temperatures are presented along the height of the furnace at the mid cut of X-Z plane for the air-fired, OF25, OF27, and OF29 combustion cases. The inlet flow temperatures of gases in the secondary air flows and in the burner gas flows were 473 and 397 K, respectively. Once the reaction processes between PC and oxidizer gases have been started the flame temperature is progressively increased to be at a peak value in the furnace zone as follows: 1864.37 K for air-fired, 1752.0 K for OF25, 1813.3 K for OF27, and 1865.0 K for OF29.

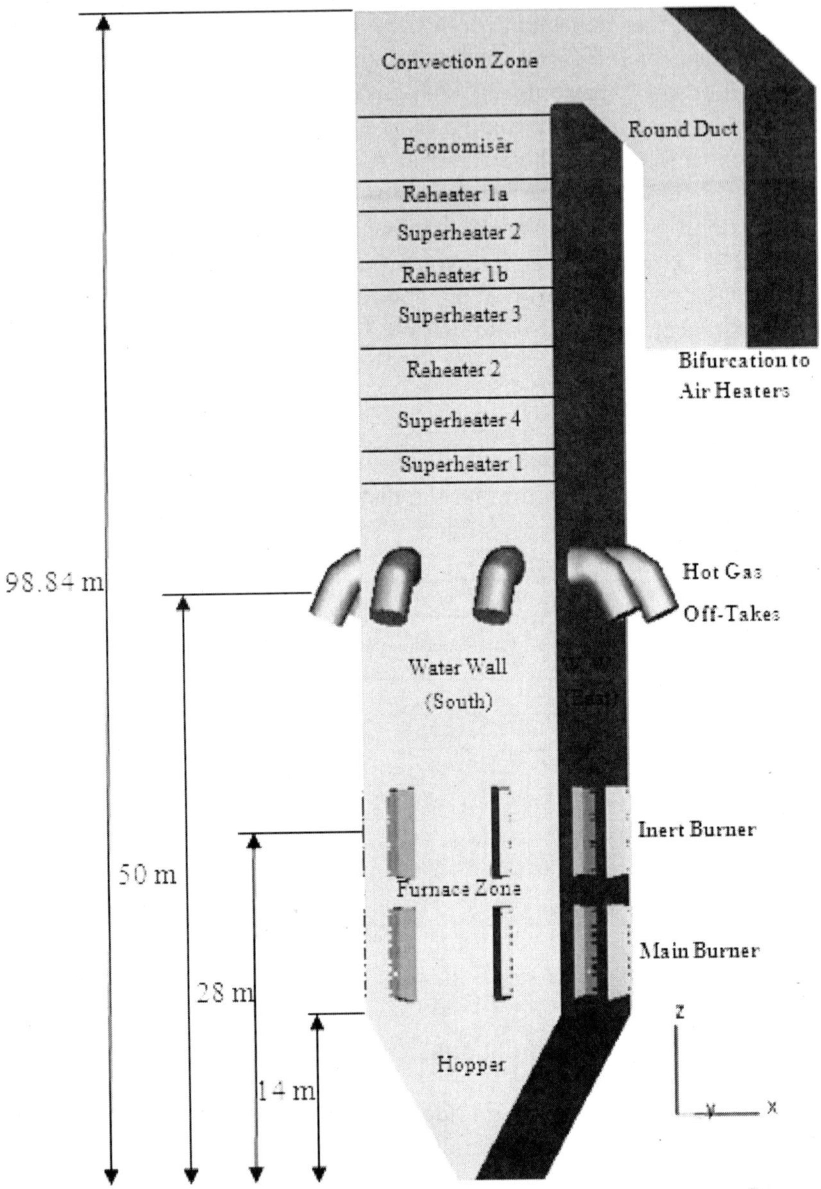

Figure 9. The geometric description of the CFD model for the boiler, unit 1 at Loy Yang A power station

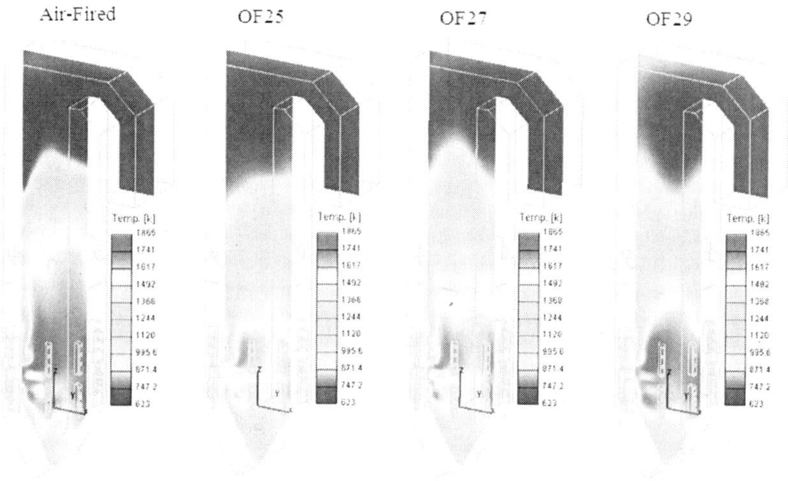

Figure 10. Distributions of the flue gas temperature (K) along the height of the furnace at the mid cut (X-Z plane) for air-fired, OF25, OF27, and OF29 combustion cases

It is clearly seen that a reduction in the levels of the gas temperature occurred when the N_2 is replaced by the CO_2 in the secondary air ducts, particularly in the OF25 and OF27cases examined. That obvious decrease in the gas temperature was mainly due to the higher volumetric heat capacity of CO_2compared to N_2 in the gas mixture. On the other hand, the maximum gas temperatures between air-fired and OF29 combustion cases were approximately identical. That improvement on the gas temperature of the latter oxy-fuel case was because of increasing O_2 concentration in the feed oxidizer gases. Furthermore, in this study, the oxygen concentration was not only the dominant factor controlling the flame temperature inside the furnace, but also the resident time of coal combustion. However, the inlet flow fields of feed oxidizer gases (O_2/CO_2), in all oxy-fuel cases, were reduced in proportion to the volumetric flow rates by fixed ratios: 83%, 77%, and 72% for OF25, OF27, and OF29, respectively with respect to the conventional firing case. The reduction in the volumetric flow rates in O_2/CO_2 cases has given a sufficient time to burn more coal particles in the combustion zone. However, according to the Chalmers' approach [53], the reduction in the volumetric feed-gas flow rates through the secondary air ducts of the furnace is an efficient method to stabilize and increase the flame temperatures.

Figure 11 shows the distributions of carbon dioxide concentration (kg/kg) along the height of the furnace at the mid cut of the X-Z plane for all cases examined. The recycled flue gas (RFG) (mainly CO_2) used in the oxy-fuel cases has increased the CO_2 concentrations in the flue gases. The differences in the CO_2 concentrations between the conventional combustion case and the oxy-fuel cases are evident due to adopting the Chalmers' approach in this study. Around five times higher CO_2 is achieved for all oxy-fuel combustion scenarios compared to the air-fired case. This increase of the CO_2 concentrations was also observed in the combined studies of the experimental investigations of Andersson (2007) [53] and numerical modelling of Al-Abbas *et al.* (2011) and 2012 [10, 51] which was conducted on the lab-scale 100 kW firing lignite furnace. The concentrations of CO_2 mass fractions at the furnace exit were equal to 18.84, 85.76, 85.01, and 84.18 wt% for the air-fired, OF25, OF27, and OF29, respectively. Due to the higher capability of carbon dioxide to absorb the combustion heat, this elevated CO_2 in the oxy-fuel cases can potentially increase the protection of the furnace wall against the hot flue gases. However, the heat transfer to the furnace wall for the air-fired and retrofitted OF29 combustion case is technically conformed through two intrinsic aspects: First the gas temperature distributions (in the furnace zone and at furnace exit, see Table 4), and secondly the wall heat flux in difference furnace wall locations.

6. CONCLUSION

The oxy-fuel combustion mode can potentially provide significant opportunities for near-zero emissions from the existing and newly-built PC power plants in comparson with the other CO_2 capture technologies such as pre-combustion and post-combustion. In this chapter a computational fluid dynamics (CFD) tool has been developed and validated to numerical turbulent combustion models of gaseous and solid fuels in different furnaces. The numerical investigations of the air-fired and several oxy-fuel combustion environments have been carried out based on the available experimental data. The observations of this research project can provide useful information for improving the combustion characteristics and heat transfer performance of the tangentially-fired boiler under oxy-fuel combustion conditions.

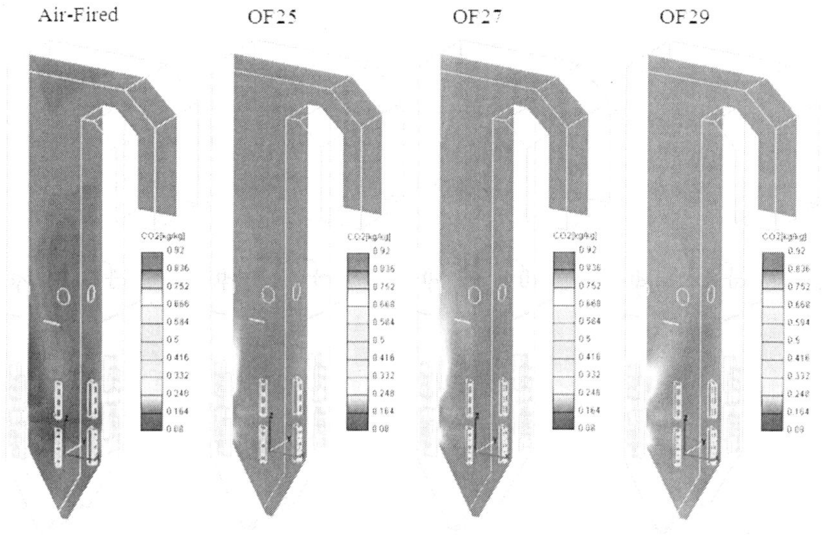

Figure 11. Distributions of carbon dioxide concentration (kg/kg) along the height of the furnace at the mid cut for all cases investigated

Table 4. Main predicted CFD results of all combustion cases examined at the final exit plane, and the net radiative heat flux on the furnace wall

Combustion Media	Temp. (K)	CO_2 (wt %)	Net Rad. Heat Flux (MW)	Carbon in Fly-Ash (%)	NO_x (ppm, dry)
Air-Fired (conventional)	635.35	18.84	778.76	2.05	218.58
Oxy-Fuel (OF25)	623.81	85.76	534.3	9.74	96.31
Oxy-Fuel (OF27)	628.27	85.01	596.37	7.04	103.2
Oxy-Fuel (OF29)	631.4	84.18	685.7	5.49	120.0

The first objective was to conduct a three dimensional numerical simulation of pulverized dry lignite in a 100 kW test facility in order to provide the confidence to predict the combustion of coal particles in oxy-fuel conditions. The purpose of this study was to simulate four different combustion environments (air-fired, OF25, OF27, and OF29) and to investigate the gas temperature distributions, species concentrations (CO_2,

H_2O, and O_2), velocity, and particles consumed in the furnace. The second research objective was to simulate the Victorian brown coal combustion in a 550 MW tangentially-fired boiler under different combustion media. This study focussed on the numerical investigation of a large-scale oxy-fuel furnace. The available power plant data (Staples and Marshall 2010) were used to validate the predicted results under the air-fired condition. The Chalmers' approach in a 100 kW facility unit (Andersson 2007) was selected to the present large-scale furnace in three O_2/CO_2 mixture conditions, referred to as OF25, OF27, and OF29, in terms of aerodynamic conditions and thermodynamic properties.

The findings confirmed that the combustion characteristics and heat transfer processes of oxy-fuel-fired cases can essentially be matched with the conventional combustion characteristics if the appropriate amounts of oxygen and RFG are to be optimised. Compared to the OF25 and OF27 combustion scenarios in the large-scale furnace, the OF29 case showed similar gas temperature levels and radiative heat transfer to that of the air-fired case. This was due to increased residence time of coal particles and O_2 concentrations in the gas mixture. It was also noticed that the higher CO_2concentrations in the oxy-fuel cases significantly affected the pyrolysis process of coal particles and thus resulted in an increase in the carbon in fly-ash. A remarkable decrease in the NO_x formation was observed because of the elimination of thermal NO process from the oxy-fuel combustion scenarios, as well as low nitrogen content and higher H_2O concentrations in the raw brown coal used. Finally, it can be concluded that the aerodynamic and thermodynamic conditions of OF29 combustion case were favourable, and closely matched the conventional combustion characteristics in several important areas in comparison with the other oxy-fuel-fired conditions.

NAN. ACKNOWLEDGEMENTS

This research program was carried out at Swinburne University of Technology in Melbourne, Australia under the sponsorship of the Iraqi Ministry of Higher Education and Scientific Research (www.mohesr.gov.iq) for developments and innovations in a clean energy sector.

REFERENCES

1. Wall, T.; Liu, Y.; Spero, C.; Elliott, L.; Khare, S.; Rathnam, R.; Zeenathal, F.; Moghtaderi, B.; Buhre, B.; Sheng, C.; Gupta, R.; Yamada, T.; Makino, K.; Yu, J., An overview on oxyfuel coal combustion--State of the art research and technology development. *Chemical Engineering Research and Design* 2009; *87* (8), 1003-1016.

2. Davison, J., Performance and costs of power plants with capture and storage of CO2. *Energy* 2007; *32* (7), 1163-1176.

3. Buhre, B. J. P.; Elliott, L. K.; Sheng, C. D.; Gupta, R. P.; Wall, T. F., Oxy-fuel combustion technology for coal-fired power generation. *Progress in Energy and Combustion Science* 2005; *31* (4), 283-307.

4. Vitalis, B. In *Overview of oxy-combustion technology for utility coal-fired boilers*, Advances in Materials Technology for Fossil Power Plants - Proceedings from the 5th International Conference, Marco Island, FL, Marco Island, FL, 2008; pp 968-981.

5. Amelio, M.; Morrone, P.; Gallucci, F.; Basile, A., Integrated gasification gas combined cycle plant with membrane reactors: Technological and economical analysis. *Energy Conversion and Management* 2007; *48* (10), 2680-2693.

6. Ratafia-Brown J; Manfredo L; Hoffmann J; M., R., Major environmental aspects of gasification-based power generation technologies.*SAIC,* 2002,*http://www.netl.doe.gov/technologies/coalpower/cctc/cctdp/bibliography/misc/pdfs/igcc/SAIC_Env_Rpt.pdf*(accessed 24 July 2010).

7. Kanniche, M.; Gros-Bonnivard, R.; Jaud, P.; Valle-Marcos, J.; Amann, J.-M.; Bouallou, C., Pre-combustion, post-combustion and oxy-combustion in thermal power plant for CO2 capture. *Applied Thermal Engineering* 2009; *30* (1), 53-62.

8. Romeo, L. M.; Bolea, I.; Escosa, J. M., Integration of power plant and amine scrubbing to reduce CO2 capture costs. *Applied Thermal Engineering*2008; *28* (8-9), 1039-1046.

9. Van Loo, S.; van Elk, E. P.; Versteeg, G. F., The removal of carbon dioxide with activated solutions of methyl-diethanol-amine. *Journal of Petroleum Science and Engineering* 2007;*55* (1-2), 135-145.

10. Al-Abbas, A. H.; Naser, J.; Dodds, D., CFD modelling of air-fired and oxy-fuel combustion of lignite in a 100 KW furnace. *Fuel* 2011; *90* (5), 1778-1795.

11. Al-Abbas, A. H.; Naser, J., Numerical Study of One Air-Fired and Two Oxy-Fuel Combustion Cases of Propane in a 100 kW Furnace. *Energy & Fuels*2012; *26* (2), 952-967.

12. Nikolopoulos, N.; Nikolopoulos, A.; Karampinis, E.; Grammelis, P.; Kakaras, E., Numerical investigation of the oxy-fuel combustion in large scale boilers adopting the ECO-Scrub technology. *Fuel* 2011; *90* (1), 198-214.

13. Andersson, K.; Johnsson, F., Process evaluation of an 865 MWe lignite fired O2/CO2 power plant. *Energy Conversion and Management* 2006; *47*(18-19), 3487-3498.

14. Andersson, K.; Johansson, R.; Johnsson, F.; Leckner, B., Radiation intensity of propane-fired oxy-fuel flames: Implications for soot formation. *Energy and Fuels* 2008; *22* (3), 1535-1541.

15. Moore, J. D.; Kuo, K. K., Effect of Switching Methane/Oxygen Reactants in a Coaxial Injector on the Stability of Non-Premixed Flames. *Combustion Science and Technology* 2008; *180* (3), 401 - 417.

16. Hjärtstam, S.; Andersson, K.; Johnsson, F.; Leckner, B., Combustion characteristics of lignite-fired oxy-fuel flames. *Fuel* 2009; *88* (11), 2216-2224.

17. Zhang, L.; Binner, E.; Qiao, Y.; Li, C.-Z., In situ diagnostics of Victorian brown coal combustion in O2/N2 and O2/CO2 mixtures in drop-tube furnace. *Fuel* 2010; *89* (10), 2703-2712.

18. Croiset, E.; Thambimuthu, K. V., NOx and SO2 emissions from O2/CO2 recycle coal combustion. *Fuel* 2001; *80* (14), 2117-2121.

19. Liu, H.; Zailani, R.; Gibbs, B. M., Comparisons of pulverized coal combustion in air and in mixtures of O 2/CO2. *Fuel* 2005; *84* (7-8), 833-840.

20. Krishnamoorthy, G.; Sami, M.; Orsino, S.; Perera, A.; Shahnam, M.; Huckaby, E. D. In *Radiation modeling in oxy-fuel combustion scenarios,*

American Society of Mechanical Engineers, Power Division (Publication) PWR, Albuquerque, NM, Albuquerque, NM, 2009; pp 615-621.

21. Yin, C.; Rosendahl, L. A.; Kær, S. K., Chemistry and radiation in oxy-fuel combustion: A computational fluid dynamics modeling study. *Fuel* 2011;*90*(7), 2519-2529.

22. Chui, E. H.; Douglas, M. A.; Tan, Y., Modeling of oxy-fuel combustion for a western Canadian sub-bituminous coal[small star, filled]. *Fuel* 2003; *82*(10), 1201-1210.

23. Murphy, J. J.; Shaddix, C. R., Combustion kinetics of coal chars in oxygen-enriched environments. *Combustion and Flame* 2006; *144* (4), 710-729.

24. Tan, Y.; Croiset, E.; Douglas, M. A.; Thambimuthu, K. V., Combustion characteristics of coal in a mixture of oxygen and recycled flue gas. *Fuel* 2006;*85* (4), 507-512.

25. Cao, H.; Sun, S.; Liu, Y.; Wall, T. F., Computational fluid dynamics modeling of NOx reduction mechanism in oxy-fuel combustion. *Energy and Fuels* 2010; *24* (1), 131-135.

26. Qiao, Y.; Zhang, L.; Binner, E.; Xu, M.; Li, C.-Z., An investigation of the causes of the difference in coal particle ignition temperature between combustion in air and in O2/CO2. *Fuel* 2010; *89* (11), 3381-3387.

27. Zhang, L.; Binner, E.; Qiao, Y.; Li, C. Z., High-speed camera observation of coal combustion in air and O2/CO2mixtures and measurement of burning coal particle velocity. *Energy and Fuels* 2010; *24* (1), 29-37.

28. Breussin, F.; Lallemant, N.; Weber, R., Computing of oxy-natural gas flames using both a global combustion scheme and a chemical equilibrium procedure. *Combustion Science and Technology* 2000; *160* (1-6), 369-397.

29. Brink, A.; Kilpinen, P.; Hupa, M.; Kjaeldman, L., Study of alternative descriptions of methane oxidation for CFD modeling of turbulent combustors.*Combustion Science and Technology* 1999; *141* (1), 59-81.

30. Venuturumilli, R.; Chen, L.-D., Comparison of four-step reduced mechanism and starting mechanism for methane diffusion flames. *Fuel* 2009; *88*(8), 1435-1443.

31. Bejarano, P. A.; Levendis, Y. A., Single-coal-particle combustion in O2/N2 and O2/CO2 environments. *Combustion and Flame* 2008; *153* (1-2), 270-287.

32. Li, Q.; Zhao, C.; Chen, X.; Wu, W.; Li, Y., Comparison of pulverized coal combustion in air and in O2/CO2 mixtures by thermo-gravimetric analysis.*Journal of Analytical and Applied Pyrolysis* 2009; *85* (1-2), 521-528.

33. Sturgeon, D. W.; Cameron, E. D.; Fitzgerald, F. D., Demonstration of an oxyfuel combustion system. *Energy Procedia* 2009; *1* (1), 471-478.

34. IEA 2009, Oxy-fuel Network, *http://www.co2captureand* (accessed 24 July 2009).

35. Zhou, W.; Moyeda, D., Process Evaluation of Oxy-fuel Combustion with Flue Gas Recycle in a Conventional Utility Boiler. *Energy & Fuels* 2010; *24*(3), 2162-2169.

36. Law, C. K., Combustion Physics. Cambridge University Press: 2006.

37. Arias, B.; Pevida, C.; Rubiera, F.; Pis, J. J., Effect of biomass blending on coal ignition and burnout during oxy-fuel combustion. *Fuel* 2008; *87* (12), 2753-2759.

38. Molina, A.; Shaddix, C. R., Ignition and devolatilization of pulverized bituminous coal particles during oxygen/carbon dioxide coal combustion.*Proceedings of the Combustion Institute* 2007; *31* (2), 1905-1912.

39. Rathnam, R. K.; Elliott, L. K.; Wall, T. F.; Liu, Y.; Moghtaderi, B., Differences in reactivity of pulverised coal in air (O2/N2) and oxy-fuel (O2/CO2) conditions. *Fuel Processing Technology* 2009; *90* (6), 797-802.

40. Kim, G.; Kim, Y.; Joo, Y.-J., Conditional Moment Closure for Modeling Combustion Processes and Structure of Oxy-Natural Gas Flame. *Energy & Fuels* 2009; *23* (9), 4370-4377.

41. Williams, F. A., A review of flame extinction. *Fire Safety Journal* 1981; *3* (3), 163-175.

42. Williams, F. A., Progress in knowledge of flamelet structure and extinction. *Progress in Energy and Combustion Science* 2000; *26* (4-6), 657-682.

43. Hurt, R. H.; Calo, J. M., Semi-global intrinsic kinetics for char combustion modeling. *Combustion and Flame* 2001; *125* (3), 1138-1149.

44. Hecht, E. S.; Shaddix, C. R.; Molina, A.; Haynes, B. S., Effect of CO2 gasification reaction on oxy-combustion of pulverized coal char. *Proceedings of the Combustion Institute* 2010; *33* (2), 1699-1706.

45. Andersson, K.; Johansson, R.; Hjärtstam, S.; Johnsson, F.; Leckner, B., Radiation intensity of lignite-fired oxy-fuel flames. *Experimental Thermal and Fluid Science* 2008; *33* (1), 67-76.

46. Huang, X.; Jiang, X.; Han, X.; Wang, H., Combustion characteristics of fine- and micro-pulverized coal in the mixture of O2/CO2. *Energy and Fuels*2008; *22* (6), 3756-3762.

47. Suda, T.; Masuko, K.; Sato, J.; Yamamoto, A.; Okazaki, K., Effect of carbon dioxide on flame propagation of pulverized coal clouds in CO2/O2 combustion. *Fuel* 2007; *86* (12-13), 2008-2015.

48. Xiumin, J.; Chuguang, Z.; Jianrong, Q.; Jubin, L.; Dechang, L., Combustion characteristics of super fine pulverized coal particles. *Energy and Fuels*2001; *15* (5), 1100-1102.

49. Smart, J.; Lu, G.; Yan, Y.; Riley, G., Characterisation of an oxy-coal flame through digital imaging. *Combustion and Flame* 2010; *157* (6), 1132-1139.

50. Rehfeldt, S.; Kuhr, C.; Schiffer, F.-P.; Weckes, P.; Bergins, C., First test results of Oxyfuel combustion with Hitachi's DST-burner at Vattenfall's 30 MWth Pilot Plant at Schwarze Pumpe. *Energy Procedia* 2011; *4* (0), 1002-1009.

51. Al-Abbas, A. H.; Naser, J., Effect of Chemical Reaction Mechanisms and NOx Modeling on Air-Fired and Oxy-Fuel Combustion of Lignite in a 100-kW Furnace. *Energy & Fuels* 2012; *26* (6), 3329-3348.

52. Al-Abbas, A. H.; Naser, J.; Dodds, D., CFD modelling of air-fired and oxy-fuel combustion in a large-scale furnace at Loy Yang A brown coal power station. *Fuel* 2012; vol. 102, pp. 646-665.

53. Andersson, K. Characterization of oxy-fuel flames - their composition, temperature and radiation. Thesis, Chalmers University of Technology, Göteborg, 2007.

54. Chui, E. H.; Majeski, A. J.; Douglas, M. A.; Tan, Y.; Thambimuthu, K. V., Numerical investigation of oxy-coal combustion to evaluate burner and combustor design concepts. *Energy* 2004; *29* (9-10), 1285-1296.

55. Dodds, D.; Naser, J.; Staples, J.; Black, C.; Marshall, L.; Nightingale, V., Experimental and numerical study of the pulverised-fuel distribution in the mill-duct system of the Loy Yang B lignite fuelled power station. *Powder Technology* 2011; *207* (1-3), 257-269.

56. Achim, D.; Naser, J.; Morsi, Y. S.; Pascoe, S., Numerical investigation of full scale coal combustion model of tangentially fired boiler with the effect of mill ducting. *Heat and Mass Transfer/Waerme- und Stoffuebertragung* 2009; 1-13.

57. Ahmed, S.; Naser, J., Numerical investigation to assess the possibility of utilizing a new type of mechanically thermally dewatered (MTE) coal in existing tangentially-fired furnaces. *Heat and Mass Transfer/Waerme- und Stoffuebertragung* 2011; *47*, 457-469.

58. Staples J.; Marshall L. Herman Resource Laboratories (HRL) Technology: Mulgrave, Victoria Australia, *www.hrl.com.au/*. Report No. HLC/ 2010/105.

Chapter 9

Exergy Analysis and Efficiency Improvement of a Coal Fired Thermal Power Plant in Queensland

R. Mahamud[1], M.M.K. Khan[1], M.G. Rasul[1] and M.G. Leinster[1]

[1] Central Queensland University, School of Engineering and Built Environment, Rockhampton, Queensland, Australia

1. INTRODUCTION

Energy security and CO_2 emission reduction are two major concerns of today's world. Improving efficiency of the energy systems is an essential option for the security of future energy and the reduction of CO_2 emissions. With the growing prosperity of the civilization, our consumption of energy is growing very rapidly. Fossil fuels remain the world's dominant primary energy supply, with its use as a versatile primary source of energy in power generation, transport and industry. However, we have finite sources of these non renewable fossil fuels and we are consuming them at a rate that cannot be sustained, leading to the risk on energy security in the future. The Intergovernmental Panel on Climate Change (IPCC), in its Fourth Assessment Report [3], identified carbon dioxide (CO_2) emissions from burning of fossil fuels as the primary contributor to climate change. Therefore, the prudent use of energy is imperative and the importance of improving energy efficiency is now very well realized by all.

Improvement of energy efficiency of power generation plants leads to lower cost of electricity, and thus is an attractive option. According to IEA (IEA 2008a), the average efficiency of coal-fired power generators of the Organisation for Economic Co-operation and Development (OECD) member[1] -countries over the 2001 to 2005 period in public sector is 37%. According to this report, the highest average efficiency of coal-fired power plants is observed in Denmark which is 43% and in United States 36%. The average energy efficiency of Australian coal-fired power plants is one of the lowest among the OECD countries which is 33%. Therefore, improving energy efficiency of coal-fired power plant in Australia is very important.

The energy conversion in a coal-fired power plant is dominantly a thermodynamic process. The improvement of energy efficiency in a thermodynamic process generally depends on energy analysis of the process which identifies measures required to be addressed. The conventional method of energy analysis is based on first law of thermodynamics which focuses on conservation of energy. The limitation with this analysis is that it does not take into account properties of the system environment, or degradation of the energy quality through dissipative processes. In other words, it does not characterize the irreversibility of the system. Moreover, the first law analysis often casts misleading impressions about the performance of an energy conversion device [4-6]. Achieving higher efficiency, therefore, warrants a higher order analysis based on the second law of thermodynamics as this enables us to identify the major sources of loss, and shows avenues for performance improvement [7]. Exergy analysis characterizes the work potential of a system with reference to the environment which can be defined as the maximum theoretical work that can be obtained from a system when its state is brought to the reference or "dead state" (standard atmospheric conditions). The main purpose of exergy analysis is to identify where exergy is destroyed. This destruction of exergy in a process is proportional to the entropy generation in it, which accounts for the inefficiencies due to irreversibility.

This research conducts exergy analysis in one unit of a coal-fired power plant in Central Queensland, Australia as a case study. The exergy analysis identifies where and how much exergy is destroyed in the system and its components. Based on the analysis, it assesses and discusses different options to improve the efficiency of the system.

2. PROCESS DESCRIPTION OF A COAL-FIRED POWER PLANT

A coal-fired power plant burns coal to produce electricity. In a typical coal-fired plant, there are pulverisers to mill the coal to a fine powder for burning in a combustion chamber of the boiler. The heat produced from the burning of the coal generates steam at high temperature and pressure. The high-pressure steam from the boiler impinges on a number of sets of blades in the turbine. This produces mechanical shaft rotation resulting in electricity generation in the alternator based of Faraday's principle of electromagnetic induction. The exhaust steam from the turbine is then condensed and pumped back into the boiler to repeat the cycle. This description is very basic, and in practice, the cycle is much more complex and incorporates many refinements.

A typical coal plant schematic is presented in Figure 1. It shows that the turbine of the power plant has three stages: high-pressure, intermediate-pressure and low-pressure stages. The exhaust steam from the high-pressure turbine is reheated in the boiler and fed to the intermediate-pressure turbine. This increases the temperature of the steam fed to the intermediate-pressure turbine and increases the power output of the subsequent stages of the turbine. Steam from different stages of the turbine is extracted and used for boiler feed water heating. This is regenerative feed water heating, typically known as regeneration. The improvement of the thermal performance of the power generation cycle with reheat and regeneration is a trade-off between work output and heat addition [8] and it can be evaluated through the efficiency of the power generation cycle.

In a typical pulverised coal power plant, there are three main functional blocks as shown in Figure 1. They are (1) the boiler; (2) the turbo-generator and (3) the flue gas clean up. The boiler burns coal to generate steam. The combustion chamber of the boiler is connected with the coal pulverisers and air supply. The water pre-heater (also known as the economiser), the super heater and the reheater are all included in this block. The steam produced in the boiler is used in the turbine as shown in Figure 1. The generator is coupled with the turbine where mechanical shaft rotation of the turbine is converted into electrical power and supplied to the power distribution grid through a transformer. The purpose of the transformer is to step up the voltage of the generated power to a level suitable for long distance transmission. The steam leaving the turbine is condensed in the condenser as shown in the Figure 1using cooling water which discharges low temperature heat to the environment. The condensate produced is pumped back to the boiler after heating through the feed water heaters. The feed water heaters use regenerative steam extracted from the turbine.

The burning of coal in the boiler of a power plant produces flue gas. The main constituents the of flue gas are nitrogen (N_2), carbon dioxide (CO_2) and water (H_2O). It carries particulate matter (PM) and other pollutants. There are traces of some oxides such as oxides of sulphur (SOx) and oxides of nitrogen (NOx) depending on the combustion technology and fuel used. The flue gas clean-up block comprises all the equipment needed for treating the flue gas. The power plant shown in Figure 1includes a DeNOx plant for NOx removal, followed by electrostatic precipitation (ESP) to remove particulate matter (PM), and wet flue gas desulfurisation (FGD) to remove SOx from the flue gas. An air-preheating unit is situated between the DeNOx and the electrostatic precipitator (ESP). There is a significant amount of heat energy leaving through the flue gas, some of which is recovered by using the air preheater. This improves the thermal performance of the process.

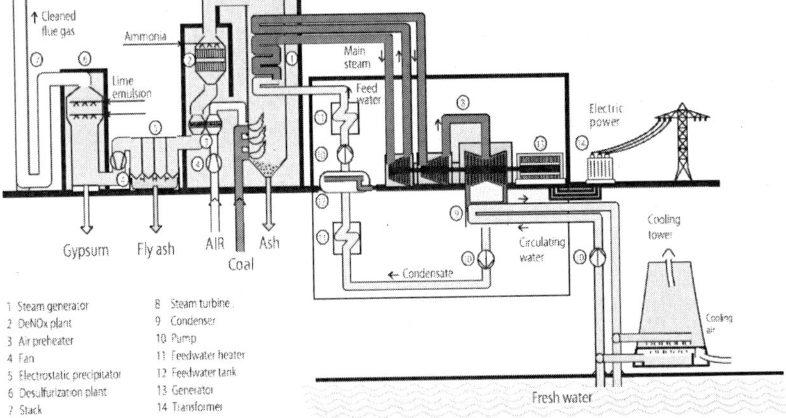

Figure 1. A typical Coal Power Plant

The properties of the coal used in the boiler and the environmental legislation and/or environmental management policy of a plant are two major factors that determine the nature of the flue gas treatment process. In some countries, due to stringent environmental regulation, coal-fired power plants need to install denitrification plants (DeNOx) for nitrogen oxide (NOx) and flue gas desulphurisation plants (FGD) for sulphur oxide (SOx) removal [9, 10]. In Australia, the coal used has a very low sulphur content and therefore, the concentration of SOx from the burning of coal in Australia is relatively low. Dave et al. [11] report an absence of stringent regulatory requirements for limiting NOx or SOx in flue gas streams in Australia. Therefore, Australian coal plants in the past have not been required to have deNOx or deSOx equipment to clean up flue gas.

In this research, a pulverized coal-fired power plant in Central Queensland, Australia has been considered as a case study. One of the units of the said plant was used to develop a process model and to perform energy analysis. This unit has Maximum Continuous Rating (MCR) of 280 MW. It spends less that 5% of its operating time at loads greater than 260 MW. Operation of the unit is mostly in the range of 100 to 180 MW range. The unit plant is a sub-critical power plant having steam outlet pressure of 16.2 MPa. The unit/plant uses thermal coal supplied from the nearby Bowen basin.

3. PROCESS MODELING AND SIMULATION

Mathematical models are effective tools for analysing systems or processes. They can be used to develop a new system or to evaluate the performance of an existing one. Mathematical modelling is widely applied to the solution of engineering problems. Modelling usually describes a system using a set of variables and equations and sets up relationships among the variables. Mathematical models are found to be very useful in solving problems related to process energy efficiency and can be utilised for both static and dynamic systems. SysCAD [1], a process modelling software package, has been found to be very effective for analysing plants for efficiency improvement. It has been used in Australia by a number of process industries, consulting companies and universities as a tool for simulating plant. Therefore, SysCAD has been employed in this study for modelling and simulating the said coal-fired power plant.

3.1. Modelling In Syscad

SysCAD can work in both static and dynamic modes. In static mode, it can perform process balances known as ProBal. SysCAD process modelling in ProBal is illustrated in Figure 2. It shows the overall approach of developing a process model in SysCAD. A process model is generally treated as a project in SysCAD. A project may have one process flow sheet or a number of flow sheets. In a project, the flowsheets can interchange data and can be interlinked.

SysCAD uses typical graphical techniques to construct a process flow diagram (PFD). Before constructing a PFD in SysCAD flowsheet, the scope of the project and data required to perform modelling needs to be defined as shown in the Figure 2. In SysCAD there are many built-in process components known as unit models. The components in a process are represented by the unit models. For modelling purposes, these unit models need to be configured based on the system requirements and performance data of the individual components used in the process as shown in Figure 2. All the unit models need to be connected appropriately to construct the PFD in a flowsheet. There are chemical species defined in SysCAD, which are used to calculate physical and chemical properties. The user can also define process components and chemical species as required if they are not available in the SysCAD component library or species database. Chemical reactions are important to perform some process modelling. SysCAD has built in features to define and simulate chemical reactions. Modelling with chemical reactions requires defining all chemical reactions in a reaction editor. The extent of a reaction based on a certain reactant can be provided as a fraction. If multiple reactions are required to define a model, the sequence of each reaction can be provided. SysCAD uses a user-defined sequence during simulation.

As shown in Figure 2, SysCAD ProBal function provides a mass and energy balance of a process and its components. The mass balance incorporates all input and output streams together with any mass additions or sinks in unit operations. This balance considers changes due to reactions or phase changes. An energy balance, on the other hand, looks at the input and output streams as well as all sources of heat transfer simply via total enthalpy. The concept is that each stream has a sum of total enthalpy and if a unit operation is balanced the sum of total enthalpy at T_{in} of all respective streams equals the sum of total enthalpy at T_{out} of all respective streams.

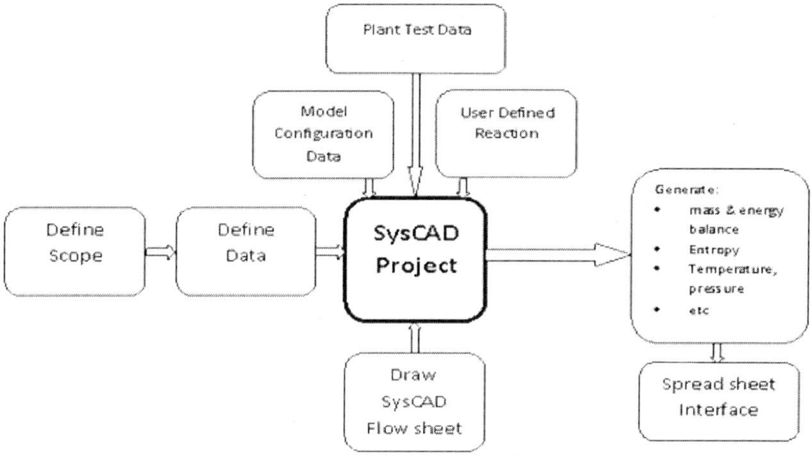

Figure 2. SysCAD Modeling [**2**]

In addition to energy and mass balances, SysCAD offers a wide range of thermo-physical data as shown in Figure 2, which is important for analysing the process and its components (e.g., temperature, pressure, entropy). SysCAD has some common process control models such as PID controller, actuator, transmitter and general controller. The general controller can be defined and used through a built-in programming language called Programmable Module (PGM). It has extended the functionality of the process modelling in SysCAD. Process data can be used in PGM to control a process and to perform calculations required for useful analysis. In addition, data from a process balance can be exported to MS Excel to perform further analysis.

In an efficiency improvement study, process energy analysis is very important for identifying where energy is lost and how effectively the loss can be minimised or recovered in the process. The results of a process modelling and simulation in SysCAD can be further analysed to observe, identify and assess energy lost in a process.

3.2. Brief Description Of Case Study

In this study, a power plant model developed based on a unit of a power plant in Central Queensland Australia. The power plant uses pulverised coal supplied through pulverisers and burnt in a boiler. The boiler of the plant is of the radiant tube type. It has natural circulation design with a low and high temperature economiser, a three-stage superheater and a two-stage single reheater. The boiler has a maximum steam outlet pressure of 16.2 MPa and a temperature of 541°C and feed water maximum temperature of 252° C.

The unit plant has a turbine to convert to convert thermal energy of steam into mechanical shaft rotation. The turbine has three pressure stages – high pressure, intermediate pressure and low pressure. In all three stages, there are stream extractors to facilitate regenerative heating of feed water heater. Three low-pressure heaters (LPH), one deaerator and two high-pressure heaters (HPH) use bled steam for regenerative feed heating. A condenser is used in the power plant to condense low pressure steam into water. The condenser is water-cooled type and it has been built for seawater operation. There is a minor loss of water in the plant process. Therefore, makeup water for the boiler feed is added into the condenser hot-well after passing through a deaerating system. It has been found that the requirement for makeup water in the boiler is very low compared to the total requirement. The condensate passes through a series of heat exchangers - LPHs, deaerator and HPHs which take heat from the regenerative bled stream as mentioned earlier.

The highest capacity of the power plant is 280 MW of electrical power. This is the maximum capacity rating (MCR) of the power plant. The capacities of all individual process components were configured with appropriate data to produce the rated power. A detailed description of the configuration of all individual process components is provided while describing the model flow sheets in the subsequent section.

3.3. Process Model Development

In this research, the power plant was represented by two separate flow sheets. They are a) Power Generation Model and b) Boiler Combustion Model. The detailed descriptions of these two model flow sheets are provided in the next sections.

3.3.1. Power Generation Model

The flow sheet for the power generation cycle is presented in Figure 3. It shows the steam cycle of the power plant. This cycle is known as the Rankine cycle [8] including reheating and regeneration.

The boiler in the power plant has feed water heater, superheater and reheater. The boiler feed water heater and super heater are included in the boiler model while the reheater is represented as a separate heater denoted as 'reheating' as shown in Figure 3.

The whole turbine is modelled using 7 unit turbine models as shown in the figure. This was done to simultaneously facilitate the use of the inbuilt SysCAD turbine model and steam extraction. For example, due to two steam extractions from intermediate pressure stage of the turbine, it is represented by two unit models namely IP_TRB1 and IP_TRB2.

The steam leaving the low-pressure turbine was connected with a condenser, which is described using a shell and tube type heat exchanger in SysCAD. The condenser is supplied with cooling water to perform steam condensation. The pressure of the steam at this stage is very low. The bled steam extracted from the turbine is recycled in to the condenser. The makeup water required in the process is added after the condenser. The condensate pump is located after the condenser, and it boosts the pressure of the condensate high enough to prevent boiling in the low-pressure feedwater heaters. The condensate mixes with the makeup water before entering the condensate pump.

There are three low-pressure heaters connected with the extracted steams from different turbine stages as shown in Figure 3. The feed water is gradually heated, taking heat from steam with increasing temperature and pressure at each stage. In a low-pressure heater, heat exchange occurs in two stages. At first, the steam condenses to its saturation temperature at steam pressure and then occurs sensible heat exchange. All the three heaters were developed based on the same principle.

The main purpose of a deaerator is to remove dissolved gases including oxygen from the feed water. Some heat exchange occurs in the deaerator. In this model, the deaerator is treated as a heat exchanging device where steam and feed water exchange heat through direct contact. The tank model built in SysCAD was used to represent the deaerator. The feed water, after heating in the deaerator, is pumped through the feed water pump. The high pressure feed water is heated through two more high-pressure heaters. The steam for the high pressure heaters is extracted from the high pressure turbine exhaust, and from an inter-stage bleed on the intermediate pressure turbine. Each of the high-pressure heaters is developed using two SysCAD heat exchange models as described earlier for the low-pressure heaters.

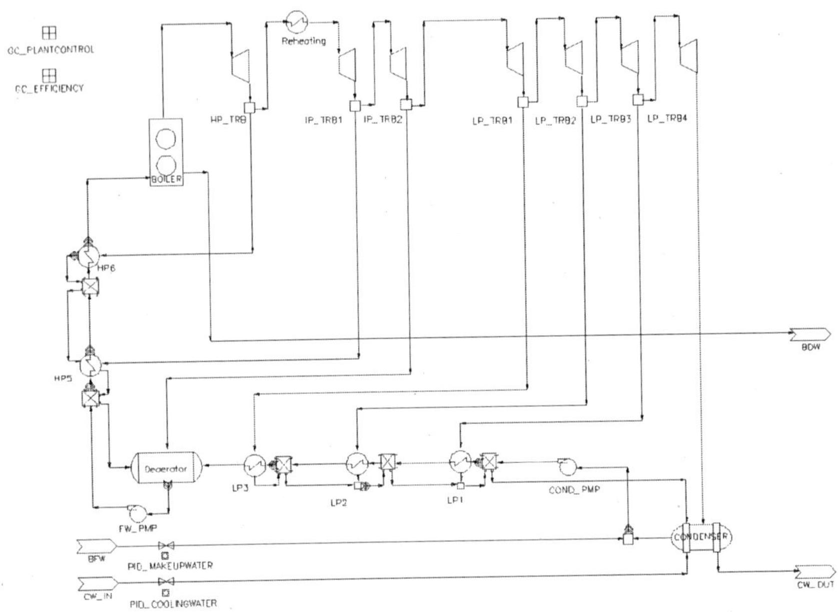

Figure 3. Power Plant Model

The operations of all the individual components used in the model flow sheet are described in detail in the subsequent discussion. The discussion includes data used to configure each component for the modelling.

3.3.1.1. Boiler and Reheating

The boiler model in SysCAD calculated energy required by the boiler based on the boiler feed water, the required drum pressure and the superheated steam conditions. It is a very simple model, which does not take into account the type of fuel used or the type of economiser. It heats the high pressure feed water stream to saturation temperature and then to the superheated condition as specified. A portion of saturated water is blown down from the boiler to discharge impurities and maintain water purity. In this research the operating conditions of the boiler were configured based on plant-supplied data presented below:

- Steam outlet pressure: 16 MPa
- Steam outlet temperature: 540 °C
- Boiler Efficiency: 90%
- Blow down: 0.5%

The high temperature steam from the boiler superheaters enters the high-pressure side of the turbine named HP_TRB. In this turbine stage, part of the steam energy is converted to mechanical shaft rotation and the pressure and temperature of the steam drops based on turbine configuration and supplied data discussed later. The steam leaving HP_TRB is taken to the reheating process where the steam is reheated to 540 °C. A portion of the steam is taken out to regeneration before it goes to reheating. The reheating model is a simple heater for sensible heating and no phase change occurs. The simple heater only calculates the heat required for reheating.

3.3.1.2. Steam Turbine

The built-in steam turbine model in SysCAD transforms steam energy into electrical power. In a flow sheet, it needs to be connected with one single steam input and one single stream output. The inlet conditions of the steam, such as temperature, pressure, mass flow and quality of steam need to be defined. Using steam inlet data and specified turbine efficiency, SysCAD calculates turbine output power and the condition of the outlet stream. The simplified energy balance calculation against a turbine is provided in the following equation:

$$W = H_{in} - H_{out} \tag{1}$$

where, W is work output of the turbine, and Hin and Hout are enthalpy of stream into and out of the turbine

The high-pressure turbine was modelled using a turbine unit model named HP_TRB. The steam exiting HP_TRB was connected with the reheating system. The reheater heats up the steam to a high temperature to improve the quality of steam. A portion of the steam is extracted before the reheater and fed to a high-pressure heater named HP6. In Figure 3, the reheated steam enters the intermediate-pressure turbine. This stage was modelled using two turbine unit models. The two models were named as IP_TRB 1 and IP_TRB 2. Bled steam from both models is taken out – from the IP_TRB 1 to the high pressure heater, HP5 and from IP_TRB 2 to the deaerator. The steam from IP_TRB2 is connected with the low-pressure turbine. The low-pressure turbine is defined using four interconnected models named LP_TRB 1, LP_TRB 2, LP_TRB 3 and LP_TRB 4. There are bled steam flows from LP_TRB 1, LP_TRB 2 and LP_TRB 3. The bled steam flows are connected with three low-pressure heaters namely, LP3, LP2 and LP1 consecutively as shown in Figure 3.

It should be noted here that the SysCAD turbine model ignored changes of potential and kinetic energy since the changes are negligible. The turbine efficiency, mechanical efficiency, outlet pressure and steam bleed of the turbine in different stages are configured with data supplied by the plant and provided in Table 1.

Table 1. Input Data for Turbine Configuration

	HP Turbine Stage(s)	IP Turbine Stage(s)		LP Turbine Stage(s)			
Turbine efficiency (%)	85						
Mechanical Efficiency (%)	98						
Outlet Pressure (kPa)	4000	1450	800	230	90	20	10
Bleed (%)	10	3	7	5	5	2	

3.3.1.3. Condenser

The condenser was represented by a shell and tube heat exchanger, and it transfers energy from one stream to another. The primary use of this model is to transfer latent heat by steam condensation. The model performed the following calculations as defined in SysCAD.

For the heat exchanger:

$$Q = UA\Delta T_{LM} \tag{2}$$

where

Q - Rate of Heat Transfer

U - Overall coefficient of Heat Transfer

A - Area available for Heat Transfer

ΔTLM- Log Mean Temperature Difference (LMTD) calculated as

$$\Delta T_{LM} = \frac{\Delta T_2 - \Delta T_1}{ln(\Delta T_2 / \Delta T_1)}$$

For Counter Current Flow $\Delta T_2 = T_{H_{in}} - T_{C_{out}}$ and $\Delta T_1 = T_{H_{out}} - T_{C_{in}}$

Where

TH_{in} – temperature of hot stream in

TH_{out} – temperature of hot stream out

TC_{in} – temperature of cold stream in

TC_{out} – temperature of cold stream out

This has been described for a counter flow heat exchanger in Figure 4.

For the heat transfer to the individual stream

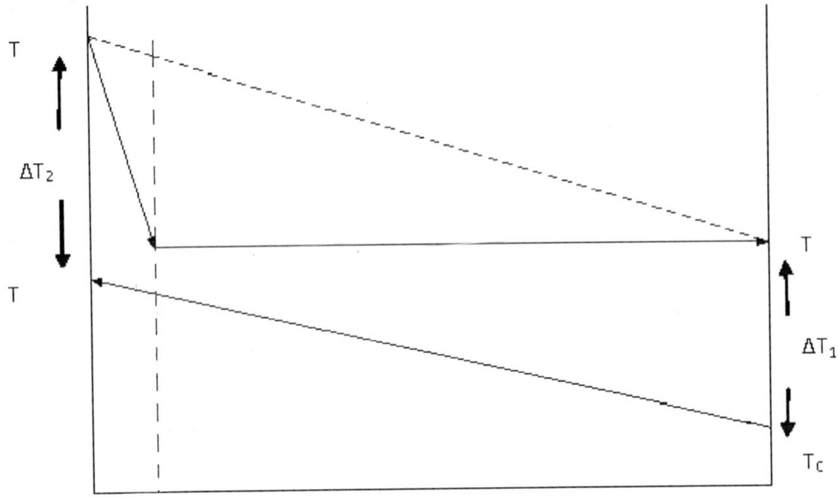

Figure 4. Log mean temperature difference [**1**]

$$Q = m(h_{in} - h_{out}) \tag{3}$$

where

Q - Rate of Heat Transfer

m - Mass flow of the stream

hin– Specific enthalpy of entering stream

hout – Specific enthalpy of leaving stream

In this model, the vapour entering the condenser first comes to the saturation temperature and is then condensed. No further cooling of the liquid occurs. The area of the heat exchanger and the cooling water required to condense the whole of the steam flow to the condenser are specified.

3.3.1.4. Low Pressure and High Pressure Heaters

The low pressure and high-pressure heaters were developed primarily using the shell and tube heat exchanger model as described earlier. Each of the heaters uses two models to achieve its desired functionality. The first one was used to condense steam into saturated water and the second exchanger was used to cool the saturated water to a temperature below the saturation temperature through sensible cooling. In SysCAD, a heat exchanger model has only one desired functionality. If one heat exchange model performs condensation of steam from some temperature above the saturation temperature, it cannot perform any further cooling to the stream. Therefore, the second heat exchanger was used to achieve the desired functionality as required by the process. All the low-pressure heaters named LP1, LP2 and LP3 and the high-pressure heaters HP5 and HP6 were developed based on the same principle.

3.3.1.5. Deaerator

The fundamental purpose of a deaerator in power generation is to remove oxygen and dissolved gases from boiler feed water. This helps prevent corrosion of metallic components from forming oxides or other chemical compounds. However, in the power generation model the deaerator was treated as a direct contact heat transfer component in order to describe it for the desired purpose of this study. In the deaerator, steam comes in direct contact with liquid water and therefore heat transfer occurs.

A tank model in SysCAD is a multipurpose model. There are sub-models available with a tank model such as reaction, environmental heat exchange, vapour liquid equilibrium, heat exchange, make-up, evaporation, and thermal split. It was used here for defining the deaerator. This tank model was configured to achieve vapour liquid equilibrium only. The other sub-models were not used. The size of the deaerator tank was kept at 10 m^3 and all the streams were brought to the lowest pressure through a built-in flashing mechanism.

3.3.1.6. Pumps

There are two pump models used in the power generation model. One is a condensate pump and the other is a feed water pump. The pump model boosts pressure of liquid to a specified pressure. In order to configure a pump in SysCAD it needs to be connected with an incoming stream and an output stream. The energy balance across a pump was achieved through using the following equation:

$$W = H_{in} - H_{out} \quad (4)$$

It was assumed that the process is adiabatic and there were negligible changes of potential and kinetic energy, and they were therefore ignored. The two pump models were configured with their required pressure boost data as provided in Table 2.

Table 2. Pump Configuration Data

Pump	Pressure Boost in kPa
Feed Water Pump	21670
Condensate Pump	1200

3.3.1.7. Pipe

The pipe model in SysCAD is used to transfer material between two units. There is a large amount of information on the stream displayed in the pipe model. It also allows some user-defined calculations using data found on the pipe. The pipe model can take pipe friction loss into consideration. However, in this project the loss in the pipe was considered to be insignificant and was therefore ignored. In the pipe model, at different points of the power generation model code was applied to perform an exergy flow calculation. This is called Model Procedure (MP) in SysCAD and the code used here is similar to the codes in PGM described earlier. The details of exergy calculation are discussed later in section 4.2.

3.3.1.8. Controls and Calculation Of Power Generation Model

There are two general controllers and two PIDs used to control the model to perform its set objectives. The general controllers were named as GC_PLANTCONTROL and GC_EFFICIENCY. GC_PLANTCONTROL was used to set the model simulation at the plant's desired rated capacity. Codes were also used to calculate the net power output and the feed water requirements. PID_MAKEUPWATER works in conjunction with GC_PLANTCONTROL to achieve the net power output set point through controlling the boiler feed water. GC_EFFICIENCY was used to calculate overall efficiency of the power plant by dividing net power output by fuel energy input rate. The net power output was calculated by deducting power used in the pumps from the generated power in the turbine. PID_COOLINGWATER was used to regulate supply of the required amount of cooling to the condenser.

3.3.2. Boiler Combustion Model

The boiler combustion model was developed to supply the desired amount of heat to the boiler. The model flow sheet is presented in Figure 5. As shown in the figure, the main components of this model are a (boiler) combustion, a water heater (economiser), a superheater, a reheater and an air preheater. The combustion model was developed using a tank model, three heaters by simple heaters and the air preheater by a heat exchanger model built in SysCAD.

3.3.2.1. (Boiler) Combustion

The combustion model used a tank model built in SysCAD to perform the transformation of chemical energy to heat energy. A reaction sub-model was configured to perform that transformation. The chemical composition of the fuel was supplied by the plant and defined in the feed of the combustor named FUEL. The reactions and their extent were defined in a reaction editor of the tank model. The chemistry of combustion is complex and depends on many different factors. It was assumed that there is sufficient air supplied to complete the combustion of coal in air. Nevertheless, the stoichiometric chemical equations used here were placed in a logical order based on the chemical affinity of the components.

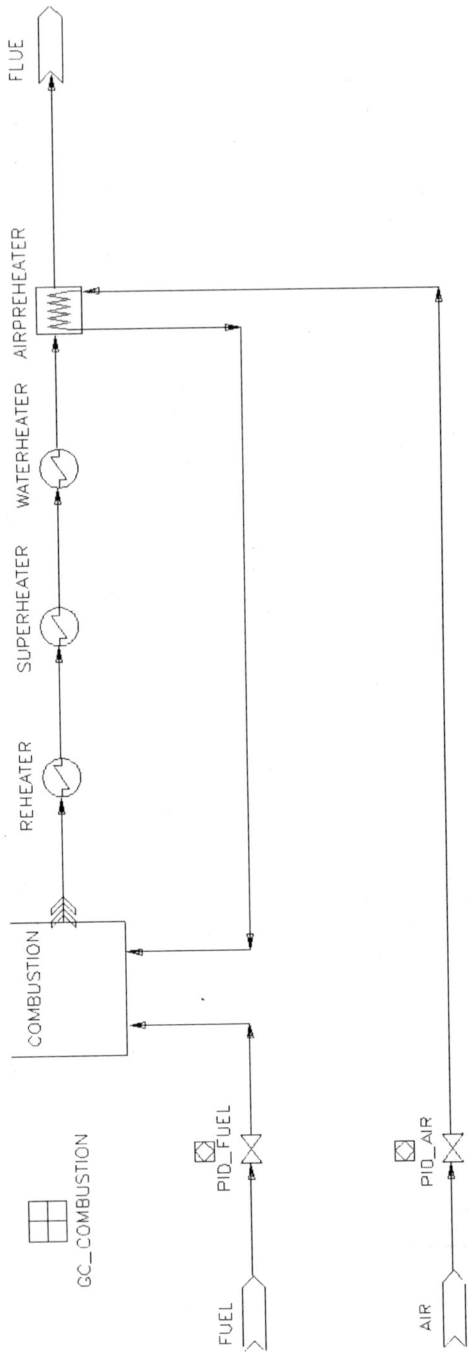

Figure 5. Boiler Combustion Model

The power plant uses thermal coal supplied from nearby coal mines. The gross calorific value (GCV) of the coal at dry ash free (daf) conditions is 30.06 MJ/kg while at air dried (ad) conditions it is 20.80 MJ/kg. Data on the composition of the coal was supplied by the power plant and is presented in Table 3.

Table 3. Coal Property Data

Proximate Analysis	Ultimate Analysis
Moisture 10.9% Ash 18.9% Volatile Materials 23.7% Fixed carbon 46.5%	Carbon 78.1% Hydrogen 3.9% Nitrogen 1.1% Sulphur 0.2% Oxygen 16.7%

As mentioned earlier, chemical reactions were performed in reaction editor in SysCAD. In the reaction editor reaction, extent and sequence is provided. The combustion reaction for this modelling purpose is provided in Table 4.

Table 4. Combustion Reaction

Reaction	Extent	Sequence
$2\ H_2(g) + 1\ O_2(g) = 2\ H_2O(g)$	Fraction $H_2(g) = 1$	1
$1\ S(s) + 1\ O_2(g) = 1\ SO_2(g)$	Fraction $S(s) = 1$	2
$1\ C(s) + 1\ O_2(g) = 1\ CO_2(g)$	Fraction $C(s) = 1$	3
$1\ H_2O(l) = 1\ H_2O(g)$	Fraction $H_2O(l) = 1$	4

Using the SysCAD property database the simulation calculates the heat of reaction (HOR) of each reaction in the combustor and then sums up all HORs to calculate overall HOR of full combustion. An environmental heat exchange is configured to allow for some heat lost to the environment from the combustor.

3.3.2.2. Water Heater (Economiser), Superheater And Reheater

These three components were modelled using the simple heater model built in SysCAD. The simple heater does not consider heating media or heater size. It only provides an estimation of heater duty required at stream outlet temperature or stream outlet temperature for specific heater duty. It can also be configured to specify heater duty irrespective of temperature. In this model, the heaters were configured to supply a specific amount of heat through heater duty calculated in the boiler and reheater in the power generation model by a so-called duty method in SysCAD. Only the duty of such heating was calculated for these heaters.

3.3.2.3. Air Preheater

The air preheater in this flowsheet was built using the heat exchanger model described previously. The process is air/air heat transfer and the heat transfer coefficient used here is 150 w/m^{2}K, which is lower than the value used for most of the liquid/liquid heat transfer.

3.3.2.4. Control And Calculation In Combustion Model

In this model, three control elements were used. One was GC_COMBUSTION, a general controller and the other two were PIDs, one for fuel and the other for air. GC_COMBUSTION calculates the amount of fuel and excess air required to complete the combustion in the combustor. The two PIDs for fuel and air regulate the required fuel and excess air to achieve a set point. The set point for excess air was 10% as supplied by the plant. The fuel requirement of the combustion was set dependent on the energy requirements in boiler and reheating in power generation model. This has increased the functionality of the model to produce any set amount of power output. Similar to the power generation model, in pipes at different points of the boiler combustion model, code was applied to perform the exergy calculation using the Model Procedure (MP) of SysCAD.

4. ENERGY ANALYSIS AND EFFICIENCY IMPROVEMENT

Energy analysis of a process is very important for identifying where energy is lost. It is performed through a process energy balance. This essentially considers all energy inputs in and outputs out of the system. When the system is balanced, the sum of all energy inputs equals the sum of all energy outputs. In a power generation plant, the objective is to convert the maximum possible energy input into useful work. According to the second law of thermodynamics, due to thermodynamic irreversibility not all energy input is converted in to useful work.

Traditionally, the energy analysis of a process is performed through energy balance based on the first law of thermodynamics. It focuses on the conservation of energy. The shortcoming of this analysis is that it does not take into account properties of the system environment, or degradation of energy quality through dissipative processes [12]. In other words, it does not take account of the irreversibility of the system. Moreover, the first law analysis often gives a misleading impression of the performance of an energy conversion device [4-6]. Getting an accurate estimate warrants a higher order analysis based on the second law of thermodynamics, as this enables us to identify the major sources of loss and shows avenues for performance improvement [7]. This essentially refers to exergy analysis that characterises the work potential of a system with reference to the environment.

4.1. Energy Balance Calculation

The energy balance was performed for the whole power generation process. As it was assumed that the energy lost in pipes is negligible, the loss of energy in process components represents the loss of energy of the whole process. The analysis was, therefore, performed to balance energy flow against all process components such as the boiler, turbine, and heat exchangers. This calculation provided information about where and how much energy is lost. The process model developed in SysCAD performs mass and energy balances considering all the input and output streams and heat and work into and out of each component. The equations used for these balances are provided in Equations 5, 6 and 7.

The mass balance for a unit process

$$\sum_{i=1}^{i} m_i = \sum_{o=1}^{o} m_o \quad (5)$$

where m is mass flow rate in Kg/s

The energy balance for a unit process

$$\sum_{i=1}^{i} E_i + \dot{Q} = \sum_{o=1}^{o} E_o + \dot{W} \quad (6)$$

where *E* is energy flow, Q˙Q˙ heat flow and W˙W˙ is work flow in Kg/s

The energy flow of the stream was calculated as

$$E = \dot{m}h \quad (7)$$

where *h* is specific enthalpy in kJ/Kg and the potential and kinetic energy of the stream are ignored.

In order to obtain a balance of energy flow against different components of the power plant Figure 6 is used. The points in these figures were chosen very carefully so that they could describe the inflow and outflow of energy carried by streams to and from each component. The work inflows and outflows were observed from the SysCAD process energy balance and were used in balance calculations where they applied.

The energy flows at the mentioned points were obtained directly from the SysCAD process balance at each mentioned point and exported to Microsoft Excel. It is important to note here that as potential and kinetic energy of the stream was ignored, SysCAD calculated the energy flow of the stream usingEquation 7. The work in and out of the components, particularly in different stages of turbine and pumps, were found in the SysCAD process balance. Therefore, the balance of energy flow was calculated observing all energy into and out of each component in the form of either heat or work. The details of energy balance calculations were performed in Microsoft Excel across various process equipment using equations provided in Table 5.

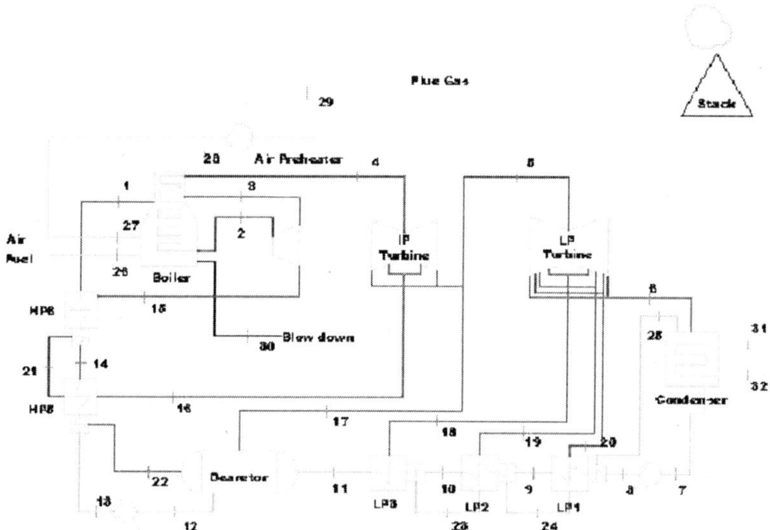

Legends

1.	High pressure boiler feed water before entering boiler	16.	IP Turbine bled stream 1
2.	Boiler superheated steam	17.	IP Turbine bled stream 2
3.	HP Turbine exhaust steam	18.	LP Turbine bled stream 1
4.	Reheated stream	19.	LP Turbine bled stream 2
5.	IP Turbine exhaust stream	20.	LP Turbine bled stream 3
6.	LP Turbine exhaust stream	21.	Condensed and cooled HP Turbine bled stream
7.	Condensate	22.	Condensed, cooled and mixed HP and IP Turbine bled stream
8.	Boiler feed water	23.	Condensed and cooled LP Turbine bled stream 1
9.	Boiler feed water after LP1	24.	Condensed, cooled and mixed LP Turbine bled stream 1 & 2
10.	Boiler feed water after LP2	25.	Recycle LP Turbine bled stream
11.	Boiler feed water after LP3	26.	Fuel for combustion
12.	Deaerated boiler feed water	27.	Preheated air
13.	High pressure boiler feed water	28.	Flue gas after combustion
14.	High pressure boiler feed water after HP5	29.	Flue gas
15.	HP Turbine bled stream	30.	Boiler blow down

Figure 6. Points of Energy and Exergy Flow in Power Plant

Here, E represents energy flow in kW. The subscripts used in the energy balance represent point numbers in Figure 6. E with subscript l represents energy lost and the corresponding process is mentioned in the subscript within the bracket. W represents work and the corresponding process component is mentioned in the subscript within the bracket.

Table 5. Energy Balance in Power plant and Capture Process

Components	Energy Balance Equation
Boiler	$E_{26}+E_{27}+E_{1}+E_{3}=E_{2}+E_{4}+E_{28}+E_{30}+E_{I(Boiler)}$
HP Turbine	$E_{2}=E_{3}+E_{15}+W_{(HP\ Turbine)}+E_{I(HP\ Turbine)}$
IP Turbine	$E_{4}=E_{5}+E_{16}+E_{17}+W_{(IP\ Turbine)}+E_{I(IP\ Turbine)}$
LP Turbine	$E_{5}=E_{6}+E_{18}+E_{19}+E_{20}+W_{(IP\ Turbine)}+E_{I(LP\ Turbine)}$
Condenser	$E_{25}+E_{6}=E_{7}+E_{I(Condenser)}$
Condensate Pump	$E_{7}+W_{(Condensate\ Pump)}=E_{8}+E_{I(Condensate\ Pump)}$
LP1	$E_{24}+E_{20}+E_{8}=E_{9}+E_{25}+E_{I(LP1)}$
LP2	$E_{23}+E_{19}+E_{9}=E_{10}+E_{24}+E_{I(LP2)}$
LP3	$E_{18}+E_{10}=E_{11}+E_{23}+E_{I(LP3)}$
Deaerator	$E_{22}+E_{17}+E_{11}=E_{12}+E_{I(Deaerator)}$
Feed Pump	$E_{12}+W_{(Feed\ Pump)}=E_{13}+E_{I(Feed\ Pump)}$
HP5	$E_{22}+E_{21}+E_{16}=E_{14}+E_{22}+E_{I(HP5)}$
HP6	$E_{14}+E_{15}=E_{21}+E_{I(HP6)}$

4.2. Exergy Balance Calculation

Exergy can be defined as 'work potential', meaning the maximum theoretical work that can be obtained from a system when its state is brought to the reference or "dead state' (under standard atmospheric conditions). The main purpose of exergy analysis is to identify where exergy is destroyed. This destruction of exergy in a process is proportional to the entropy generation in it, which accounts for the inefficiencies due to irreversibility. Exergy analysis helps in identifying the process of irreversibility leading to losses in useful work potential and thus pinpointing the areas where improvement can be sought.

Rosen [6] identified various exergy studies conducted by different researchers and found the significance of application for process energy analysis whether it is small or large but particularly more important for energy intensive ones e.g., power generation where large scale energy conversion happens. From this point of view, exergy analysis for power plants can be useful for identifying the areas where thermal efficiency can be improved. It does so by providing deep insights into the causes of irreversibility.

The exergy is thermodynamically synonymous to 'availability' of maximum theoretical work that can be done with reference to the environment. Som and Datta [13] define specific exergy, a (in kJ/Kg) in general terms as in Equation 8.

$$a = \underbrace{k + \varnothing + (u - u_r) + p_r(v - v_r) - T_r(s - s_r)}_{\text{Physical Exergy}} + \underbrace{a_{ch}}_{\text{Chemical Exergy}} \quad (8)$$

where k (in kJ/kg) is the specific kinetic energy of the system and Ø (in kJ/kg) is the potential energy per unit mass due to the presence of any conservative force field. T (in K), p (kN/m2), u (in kJ/kg), v (m3/Kg) and s (kJ/Kg-K) are the temperature, pressure, specific internal energy, specific volume and specific entropy, respectively, while a_{ch} represents the specific chemical exergy. The terms with the subscript r are the properties of the exergy reference environment. The generalised equation above can be simplified or specified based on a process. The kinetic and potential energies are small in relation to the other terms, and therefore they can be ignored.

The exergy flow was calculated at different points before and after the process components in different streams. The exergy of a stream was treated for both physical and chemical exergy. The physical exergy accounts for the maximum amount of reversible work that can be achieved when the stream of a substance is brought from its actual state to the environmental state.

According to Amrollahi et.al. [14] this can be evaluated as

$$a = (h - h_r) - T_r(s - s_r) \quad (9)$$

where h is the specific enthalpy in kJ/Kg.

This equation was applied for all liquid streams in the power plant. For gas streams this equation was reorganised in terms of specific heat of gas c_p and evaluated as

$$a = c_p(T - T_r) - c_p T_r ln\,(T / T_r) \qquad (10)$$

where c_p is the specific heat in kJ/kgK.

The reference environment mentioned earlier was considered as temperature 27.8 °C and pressure 101 kPa. At this temperature and pressure, the enthalpy and entropy were obtained for water. UsingEquation 9 exergy flow at different points of the power plant steam cycle was calculated. Equation 10was applied to the different gas streams such as the air stream entering combustion and the flue gas following combustion.

The exergy flow rate can be calculated with the following equation

$$A = ma \qquad (11)$$

where *A* is exergy flow rate in kJ/s

The fuel enters the combustion at the reference environmental condition. Therefore, it enters the boiler combustor with only the chemical exergy with it. The chemical exergy flow of the fuel was calculated through Equation 12.

$$A = \sum_{i=1}^{i} \Psi_i{}^* \Phi_i \qquad (12)$$

where *A* in kJ/s, *Ψ* is Molar flow in kmol/s and *Φ* is Molar Exergy in kJ/kmol.

Standard Molar Exergy at 298 °C and 1 atm was found in the appendix of Moran and Shapiro [8] and used in Equation 12 to calculate exergy flow with fuel. The values of molar exergy of important chemical species in standard conditions are presented in Table 5. The reference temperature is slightly higher than this temperature. At this small temperature difference, the change of molar exergy is negligible and therefore ignored.

Once the exergy flows of all input and output streams were calculated, and the work inputs or outputs were obtained from different components, the destruction of exergy could be calculated by usingEquation 13 after calculating all exergy inputs and outputs for a unit operation.

$$\sum_{i=1}^{i} A_i = \sum_{o=1}^{o} A_o + A_d \tag{13}$$

where subscripts *i, o* and *d* are input, output and destruction.

Table 6. Molar Exergy in Standard Condition

Components	Molar Exergy in (kJ/kmol)
Carbon	404590
Hydrogen	235250
Sulphur	598160
Oxygen	3950
Water Vapour	8635
Nitrogen	640

The simulation of the whole process model produces process a mass and energy balance. The simulation performed the balance at steady state using the configuration inputs provided to each component described in section 3.3. There is a wide range of thermo-physical data available in addition to process mass and energy balances. Using those data the exergy flow calculation was performed at the different points. Equations 9~12 are used to calculate specific exergy and exergy flow at different points.

The balance of exergy flow against different components of the power plant was performed individually for each component observing exergy flows into and out of the component. Equation 13 is used to calculate an exergy balance which is similar to energy balance calculation described earlier. The same points in the power plant were used to observe the exergy flow of the streams as in the energy balance calculation. Similar to the energy flow calculation, the work inflow and outflow were obtained from the SysCAD process energy balance and were used in exergy balance calculations where they applied. The details of the exergy balance calculations of the power plant are presented in Table 7. As was the case for the energy balance calculation, the results of the exergy balance were exported to Microsoft Excel where, using equations described in Table 7, the exergy balance for all individual components was performed.

Table 7. Exergy Balance in Power Plant

Components	Exergy Balance Equation
Boiler	$A_{26} + A_{27} + A_1 + A_3 = A_2 + A_4 + A_{28} + A_{30} + A_{d(Boiler)}$
HP Turbine	$A_2 = A_3 + A_{15} + W_{(HP\ Turbine)} + A_{d(HP\ Turbine)}$
IP Turbine	$A_4 = A_5 + A_{16} + A_{17} + W_{(IP\ Turbine)} + A_{d(IP\ Turbine)}$
LP Turbine	$A_5 = A_6 + A_{18} + A_{19} + A_{20} + W_{(IP\ Turbine)} + A_{d(LP\ Turbine)}$
Condenser	$A_{25} + A_6 = A_7 + A_{d(Condenser)}$
Condensate Pump	$A_7 + W_{(Condensate\ Pump)} = A_8 + A_{d(Condensate\ Pump)}$
LP1	$A_{24} + A_{20} + A_8 = A_9 + A_{25} + A_{d(LP1)}$
LP2	$A_{23} + A_{19} + A_9 = A_{10} + A_{24} + A_{d(LP2)}$
LP3	$A_{18} + A_{10} = A_{11} + A_{23} + A_{d(LP3)}$
Deaerator	$A_{22} + A_{17} + A_{11} = A_{12} + A_{d(Deaerator)}$
Feed Pump	$A_{12} + W_{(Feed\ Pump)} = A_{13} + A_{d(Feed\ Pump)}$
HP5	$A_{22} + A_{21} + A_{16} = A_{14} + A_{22} + A_{d(HP5)}$
HP6	$A_{14} + A_{15} = A_{21} + A_{d(HP6)}$

In Table 7, the notation A represents exergy flow rate in kW. The subscript number in this table represents the point number in Figure 6. A with subscript d represents exergy destruction in the corresponding process mentioned within the bracket. W represents work and the corresponding process component is mentioned in a subscript within the bracket.

5. RESULT AND DISCUSSION

As mentioned earlier, exergy analysis is performed to assess the loss of useful work potential; that is, the exergy destruction. The result of the exergy balance performed in the power plant is presented inFigure 7. In this figure, the exergy destruction is shown in percentages for different components of the power plant. The reason behind choosing percentage of total exergy destruction in different components is that these figures can easily be used to locate where the maximum exergy is destroyed. In other words, they can help direct the focus of the improvement by considering the components where most of the exergy is destroyed. The reason why this amount of exergy is destroyed is also important for identifying how best to reclaim the lost energy back in the process.

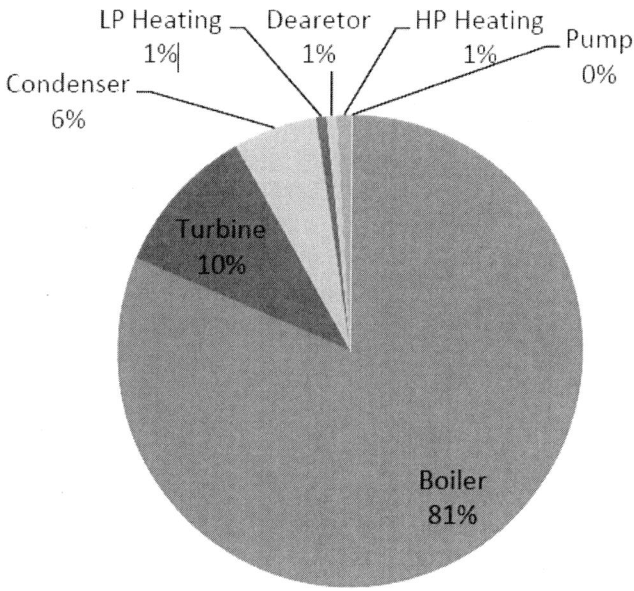

Figure 7. Exergy Destruction (%) in Power Plant

Figure 7 shows the result of this exergy destruction analysis. Most of the exergy is destroyed in the boiler, which accounts for 81% of total exergy destroyed in the power plant. It is important to note that the boiler includes both reheating of steam after expansion in the high-pressure turbine and preheating of air coming into the combustor. The second-largest source is the turbine where about 10% of the exergy is destroyed. These two components of the plant contribute more than 90% of the total exergy loss of the whole plant. The condenser is the third-largest contributor with about 6% of the exergy loss. The results of exergy analysis are markedly different from the results of the energy balance, which shows most of the energy being lost in the condenser.

A comparison of results of the energy and exergy balances in the power plant was conducted and the results are presented in Figure 8. It shows that there are very significant differences between exergy destruction and energy lost for different process components. It is important to mention here that only major components in the power plant, where most of energy or exergy loss or destruction take place, are considered in this comparison.

The energy balance showed that the primary source of energy loss is the condenser where 69% of the total loss occurs. In contrast, the exergy analysis showed that the loss from the condenser was only 6% of the total. According to the energy balance, the second-largest source of energy loss is the boiler, which accounts for 29%. However, the exergy balance revealed that the loss of useful work potential is in the boiler, with losses of more than 80%. It has been observed that there is a huge amount of energy lost in the condenser but the amount of useful energy, that is, exergy, is not very significant. In other words, it indicates that the waste heat in the condenser does not have much potential to be utilised as a source of work and to improve the efficiency the power plant. On the other hand, further investigation of the exergy lost in the boiler may show some opportunities for improvement.

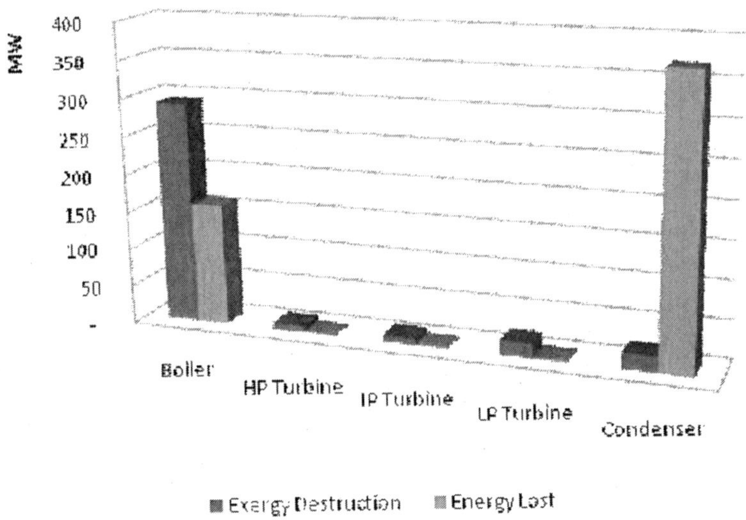

Figure 8. Exergy Destruction vs. Energy Loss in Power Plant

In an attempt to improve the efficiency of the power plant, the exergy analysis was revisited. It is found that the largest exergy destruction in the process occurs in the boiler. Therefore, it is considered first for a detailed investigation. It has been found that there are three elements which contribute to the huge boiler loss. They are 1) the boiler's internal heat transfer mechanism from the combustor to the heating medium, which determines the boiler's internal efficiency, 2) the heat loss in the departing flue gas stream and 3) the loss in the blowdown stream of the boiler. The contribution of the three losses of exergy flow in the boiler is presented in Figure 9. The simulation at steady state (280 MW electrical power output) showed that the flue gas after air preheating is leaving with an exergy flow of 4871 kW which constitutes about 2% of the total boiler exergy lost. The blowdown stream of the boiler is another source of boiler exergy destruction. It carries about 718 kW exergy flow. The boiler's internal heat transfer mechanism is responsible for most of the exergy loss occurs in the boiler.

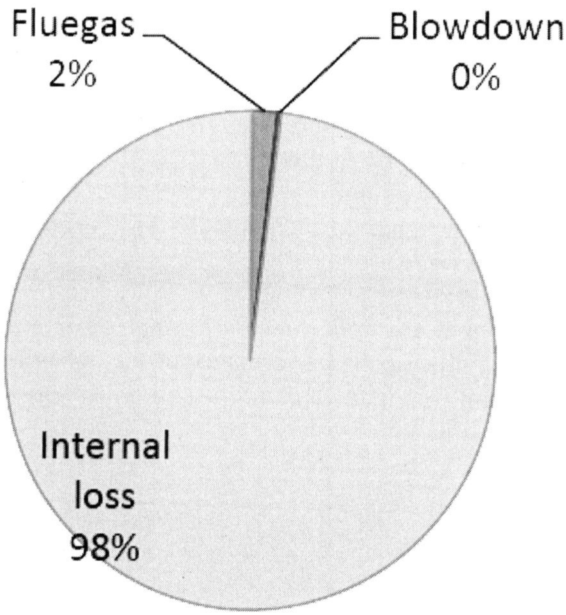

Figure 9. Boiler Exergy Losses in %

The exergy loses from flue gas and blowdown stream could be easily reutilised in the process through some heat recovery systems. However, the greatest single amount of exergy is destroyed in the boiler's internal heat transfer arrangement. If the exergy lost through the boiler could be utilised in the system, this would improve the efficiency of the power plant very significantly. The chemical reactions occur in the boiler at a temperature of 1885°C and produce huge amounts of exergy. However, due to the limitations of the material used in the boiler, it is not capable of transferring the full amount of useable heat energy or exergy to the boiler feed water which is heated to only 540°C. In modern power plants, this limitation has received much attention with the invention of new materials for heat transfer used in boilers. However, the power plant used as the subject of this study was an aging plant. The improvements needed to address such a big loss would require huge physical changes of the boiler system and may need further detailed investigation in terms of both technical and economic viability.

The exergy lost in the turbine is investigated by looking at the losses at different stages of the turbine. The exergy lost in all three turbine stages is found to be due to the turbine's internal performance. This can be better described as converting thermal energy to mechanical energy and then to electrical power. The turbine system used in the plant is highly compact and specially designed for the process. Therefore, the task of reducing exergy destruction or improving the efficiency of the turbine system is very specialised (in terms of its internal system). Similar to the boiler system improvement, it also needs further detailed investigation to assess the technical and economic feasibility of such changes.

It is noted that there are opportunities to improve energy efficiency of power plants by improving the performance of the boilers and the turbine system. The current trends towards ultra-supercritical power plant cycles are consistent with this aim.

6. CONCLUSIONS

In this study, exergy analysis of the power plant identifies areas where most of the useful energy is lost and discusses potential of the lost energy for improvement of the plant energy efficiency. It shows that the boiler of a subcritical power generation plant is the major source of useful energy lost. Only negligible amounts of useful waste energy can be recovered through implementing some heat recovery system. In order to achieve significant improvement of energy efficiency the boiler and turbine systems need to be altered, which require further techno-economic study.

NOTES

[1] - OECD members are economically developed countries.

REFERENCES

1. SysCAD*SysCAD*. 2011Available from: http://www.syscad.net/syscad.html.

2. SysCAD*Introduction to creating a SysCAD project*, K.A.P. Ltd, Editor. 2010

3. IPCC*Climate Change 2007Mitigation. Contribution of Working Group III to the Fourth Assessment Report of the Intergovernmental Panel on Climate Change*, O.R.D. B. Metz, P.R. Bosch, R. Dave, L.A. Meyer (eds), Editor. 2007, Cambridge University Press: Cambridge, United Kingdom and New York, NY, USA.

4. T. K Ray, et al*Exergy-based performance analysis for proper O&M decisions in a steam power plant.* Energy Conversion and Management, 201013331344

5. P Regulagadda, I Dincer, and G. F Naterer, *Exergy analysis of a thermal power plant with measured boiler and turbine losses.* Applied Thermal Engineering, 2010970976

6. S. K Som, and A Datta, *Thermodynamic irreversibilities and exergy balance in combustion processes.* Progress in Energy and Combustion Science, 2008351376

7. M. A Rosen, *Can exergy help us understand and address environmental concerns?* Exergy, An International Journal, 2002214217

8. M. J Moran, and H. N Shapiro, *Fundamentals of Engineering Thermodynamics*. 6 ed. 2008John Willy &Sons Inc.

9. M Bryngelsson, and M Westermark, *CO2 capture pilot test at a pressurized coal fired CHP plant.* Energy Procedia, 200914031410

10. Ö Korkmaz, G Oeljeklaus, and K Görner, *Analysis of retrofitting coal-fired power plants with carbon dioxide capture.* Energy Procedia, 200912891295

11. N Dave, et al*Impact of post combustion capture of CO2 on existing and new Australian coal-fired power plants.* Energy Procedia, 20112 0052019

12. G. P Hammond, *Industrial energy analysis, thermodynamics and sustainability.* Applied Energy, 2007675700

13. G Tsatsaronis, *Thermoeconomic analysis and optimization of energy systems.* Progress in Energy and Combustion Science, 1993227257

14. Z Amrollahi, I. S Ertesvåg, and O Bolland, *Optimized process configurations of post-combustion CO2 capture for natural-gas-fired power plant-Exergy analysis.* International Journal of Greenhouse Gas Control, 201113931405

Chapter 10

Energetic Analysis of Two Thermal Power Plants with Six and Seven Heaters

Miguel Toledo[1], Juan Abugaber[1], Raúl Lugo[2], Martín Salazar[3], Alfredo Rodríguez[4], Aldo Rueda[1]

[1]Applied Hydraulics and Thermal Engineering Laboratory, National Polytechnics Institute, Mexico City, Mexico

[2]Hydraulic Engineering and Process Department, Metropolitan Autonomous University, Mexico City, Mexico

[3]Industrial Engineering and Mechatronics Department, Tecnológico de Estudios Superiores de Ecatepec, Ecatepec de Morelos, Mexico

[4]Centro de Investigación en Ingeniería y Ciencias Aplicadas, CIICAp/UAEM, Cuernavaca, Mexico

ABSTRACT

An energetic analysis of two power plants with six and seven heaters was realized in this work and also a feedwater heater was taking into consideration. This analysis considered the temperature and pressure ranges under which power generating plants work. The unit thermal consumption (UTC), the specific fuel consumption (SFC) and the specific steam consumption (SSC) were ob-tained for power generation. The energetic analysis to the cycles of steam with regeneration with six and seven heaters was realized. There was a difference of 0.5% in the values of UTC, SFC and SSC when the pressures of condensation changed. Also an analysis of the behavior of the thermal efficiency in relation to the rise of numbers of heater from two to seven is presented.

KEYWORDS

Energetic Analysis; Feedwater Heater; Thermal Efficiency; Rankine Cycle

1. INTRODUCTION

In Mexico, most of the thermoelectrical plants work according to Rankine cycle with superheat, reheat and regeneration. Many of these plants are working with heating with six and seven heaters and one of them is feedwater heater. Generally the feedwater heater is placed on conditions of a medium pressure ranging from 10 to 15 bars and a temperature from 150°C to 200°C. Currently, 40% of the electrical power is generated in thermoelectricals. According to this, in the present work, an energetic study in thermal plants was realized in order to predict the thermal efficiency under operating conditions and some environmental variables. A thermodynamic analysis of a Rankine cycle reheat of steam power plant is conducted in terms of the first law of thermodynamic analysis (i.e. energy analysis) and the second law analysis (i.e. exergy analysis). Two heaters considering the operating parameters were used but design parameters were not considered. Steam quality at the end of the expansion are less than 0.88 [1,2]. If a thermal plant works under these conditions the steam turbine would present erosion zones.

The power generation plants have two limitations in their operation: the technological and environmental. Most thermoelectrical plants in Mexico operate in conditions of the inlet to the steam turbine of 124 bar and a temperature above 538°C, and different conditions of condensing pressure depending of the environment conditions of the place where they are installed, although the newer plants work at a pressure of 160 bar and a temperature of 600°C [3]. An energetic analysis is realized at two power plants with six and seven heaters, placing in the same order with a feedwater heater. This analysis considers the temperature and pressure ranges at which power generating plants work, to obtain the unitary heat consumption, specific consumption of steam and fuel required to generate power [4]. It also includes an analysis of how many heaters are necessary in the operation of a power plant, and in this manner, a simulation of operating plants with two heaters and the simulation of operating plants up to six heaters and one feedwater heater was realized [5]. The energetic analysis is performed using mass and energy balances in each of the units that constitute the steam cycle, which serve as the basis for a computer program in Visual Basic platform. For this analysis, the operating parameters of the Power station in Altamira, Tamaulipas are considered to operate seven heaters and the one in the city of Lerdo in Durango to operate with six heaters, both power stations generate 160 MW [6]. Likewise, the same analysis is performed in power stations that operate with less number of heaters, varying the condensing pressure and the pressure at the inlet of the steam turbine but maintaining constantly all other conditions of operation. Table 1 shows the operating conditions to which the energetic analysis is realized.

2. STEAM CYCLES

The steam cycles are analyzed in Rankine cycles with superheat, reheating and regeneration with arrangements on the train heating of 2, 3, 4, 5, 6 and 7 heaters, the evaluation is at the same operating conditions and using the optimum pressure ratio p3/p1 as design condition making sure that in the last stage of expansion in the turbine the steam has a higher quality than 0.88.

The temperature entropy diagrams of power stations with six and seven heaters are shown inFigures 1 and 2. The feedwater heater is placed in the third position on the condenser. In the following figures the six different processes of heating, superheat, reheat, regeneration, pumping and expansion of steam are shown The regeneration in the Rankine cycles has advantages because it increases the feed water temperature on its way back to the heater and, as a result, it increases the thermal efficiency of the process of feed heating and because it requires less fuel flow to generate power.

In the Rankine cycle with regeneration, the steam at high pressure and the temperature produced in the steam generator is fed into the turbine where expands its volume in each one of its stages, then part of this steam is taking out to take advantage of its high temperature to preheat the feed line of the steam generator.

In steam turbine plants, several feed heaters are introduced. The process results in an improvement in the thermal efficiency plant.

Table 1. Operating Conditions.

Design Conditions		Operating Conditions	
h_{Ts}	0.9	T_1 (°C)	538
h_{Pump}	0.8	p_1 (bar)	124.85
p_3/p_1	0.25	P_{cond} (bar)	0.11
TTD_{HP} (°C)	0	PP (MW)	160
TTD_{LP} (°C)	−2.77	LHV_{gas} (kJ/kg)	41860 kJ/kg

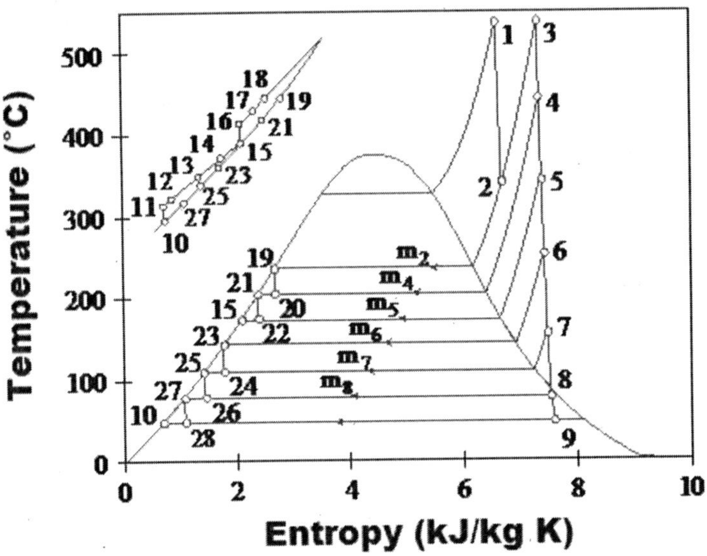

Figure 1. Diagram of temperature and entropy of regenerative cycle with six heaters.

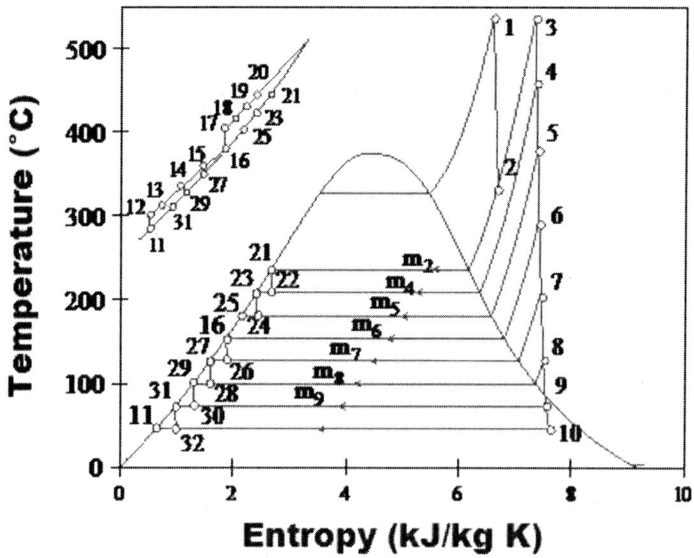

Figure 2. Diagram of temperature and entropy of regenerative cycle with seven heaters.

With the Rankine cycles, superheated, reheating and regeneration and the operating conditions ofTable 1, a parametric analysis was realized varying the condensation pressures and input to the steam turbine, getting as a result the thermal efficiency, unitary thermal consumption, the fuel flow, the specific consumption of fuel and steam, which are defined by the following equations:

Thermal efficiency

$$\eta_{th} = \frac{w_m}{q_a} \quad (1)$$

Unitary Thermal Consumption

$$UTC = \frac{3600}{\eta_{th}\eta_p\eta_e} \quad (2)$$

with $\eta_m = \eta_e$ =1 Specific Fuel Consumption

$$SFC = \frac{3600m_f}{P} \tag{3}$$

Specific Steam Consumption

$$SSC = \frac{3600m_S}{P} \tag{4}$$

where:

w_m: Rate of Work by mass unit; [kJ/kg]q_a: Rate of Heat Transfer by mass; [kJ/kg]$m_{f,s}$: Fuel and Steam Flow; [kg_c/s, kg_v/s]P: Pressure; [bar]h_e: Electrical Efficiency; [-]h_p: Mechanical Efficiency; [-].

To evaluate the above equations, thermodynamic properties per unit mass in each of the states in the steam cycle, temperature, pressure, enthalpy, entropy and volume, were found. Next, the balances of matter and energy were obtained in each of the equipment and processes involved in regenerative cycles and with these it is possible to obtain the fuel and steam flows.

3. RESULTS

According with the operation conditions of Table 1, with variation of condensation pressure and the inlet steam of the turbine, and maintaining other operating conditions constant, the following results are obtained.

In Figure 3 it is shown the steam flow of water to generate a power of 160 MW. There is a decrease in the SSC when it operates with seven heaters for steam cycles with regeneration. When it operates with 2, 3, 4 and 5 heaters this decrease is approximately of 8.3%, 3.7%, 2.2% and 1.2% respectively. However, when compared to the regeneration with six and seven heaters, SSC has a decrease of 0.4%, which means that the regeneration steam consumption decreases for power generation, because it increases the useful work and reduces the steam flow, as shown in Equation (5).

The same behavior with the SFC is shown in Figure 4, but when it operates with seven heathers the variation is slightly less than using 2, 3, 4, 5, 6 heaters, which reduces by 6.2%, 2.5%, 1.5%, 0.9% and 0.3% respectively.

This variation is more important than the previous due to the water is less expensive than the fuel burnt. Therefore, by increasing the number of heaters to eight would increase less than 0.3%, in accordance with the tendency of Figures 3 and 5:

The thermal efficiency at different pressures of condensation for steam cycles with regeneration is shown in Figure 5, it is observed that for all steam cycles a greater efficiency is achieved when the condensation pressure is 0.04 bar, likewise, it is shown that for the operating conditions the higher efficiency of the cycles is obtained when working with seven heaters (44.52%), although the increase with respect to the cycle that has six heaters (44.39%) is 0.13%.

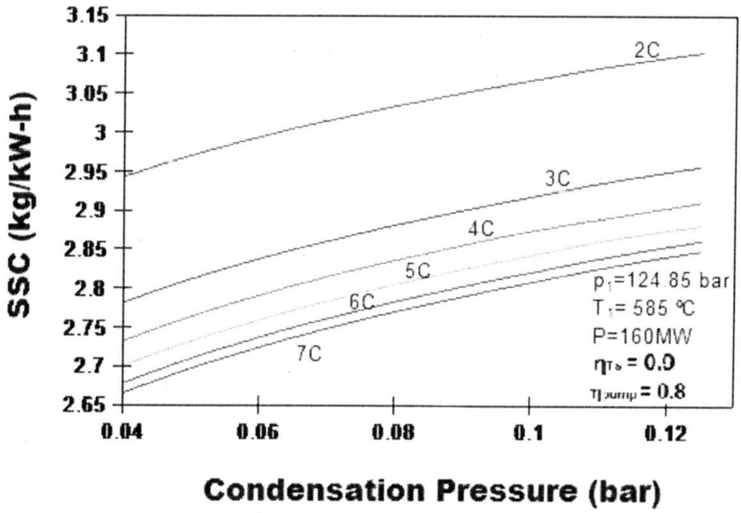

Figure 3. SSC to different pressure of condensation for steam cycles with regeneration.

The behavior of UTC at different pressures of condensation is shown in Figure 6, which shows that the difference between operating with seven or six heaters consists in providing approximately 24 kJ/kWh therefore it is efficient to work with these cycles rather than 2, 3, 4, 5 which maintain a significant difference when compared with the seven heaters, 532, 208, 123, 70 kJ/kWh respectively.

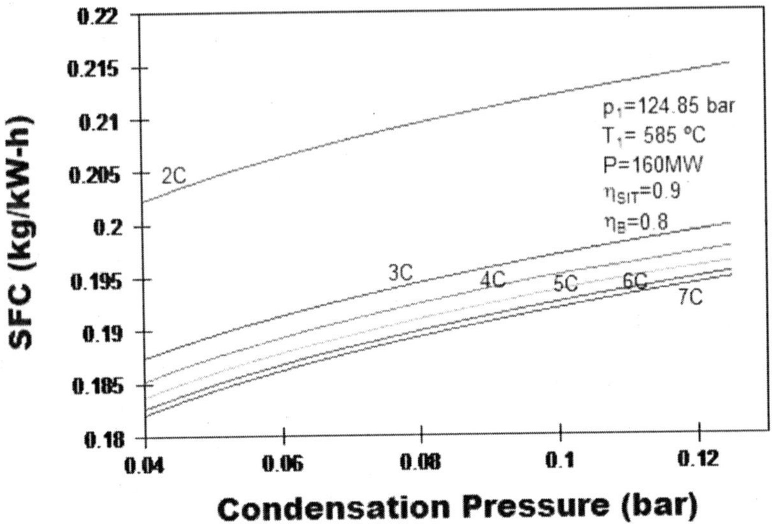

Figure 4. SFC to different pressure of condensation for steam cycles with regeneration.

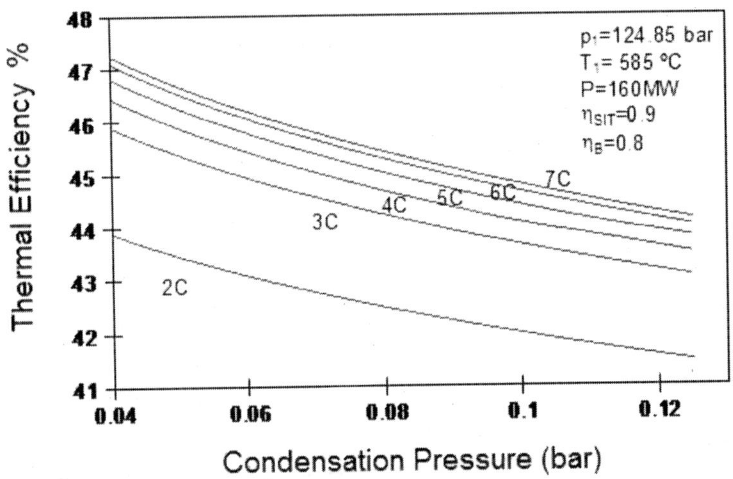

Figure 5. Thermal Efficiency to different pressures of condensation for steam cycles of regeneration.

The difference in the operation of the steam cycles with regeneration with seven or six heaters, by varying the pressure of condensation and steam, is a little increment in the SFC. The SSC UTC and condensation at different pressures and different conditions at the turbine inlet are shown in Figure 7. The conditions of point "a": the UTC is 8085.61 kJ/kW-h and the SSC of 2825 kg/kW-h, maintaining the inlet pressure to the turbine and low condensing pressure in 0.077 bar results in a decrease of the UTC and the SSC by approximately 2.2% in point "b". The same decrease of these parameters is obtained by allowing the condensation pressure constant and increasing the steam pressure from 124.85 to 160 bar, but it requires a greater amount of steam flow, as the SSC increased, to generate 160 MW.

The behavior of the thermal efficiency and the useful work at different pressures is shown in Figure 8, where the behavior is similar to Figure 7. However, operating conditions where maximum thermal efficiency does not imply that it has a maximum useful work. Considering the same points of comparison of Figure 7 it must be achieved the same thermal efficiency in the point "b" and "c"; point "b" is to be provided a larger amount of water.

The variation of the fuel flow when the operation is at different pressures of steam and condensation can be seen in Figure 9. By increasing the pressure at the inlet of the steam turbine, there is a decrease in fuel flow and the heat supplied, but this decrease is not higher than the one obtained when the condensing pressure decreases, e. g. for the conditions of the operating point to increase the pressure from 124.85 to 160 bar (points a and b) maintaining the pressure of 0.11 bar, there is a decrease of 0.18 kg/s. Now, if the pressure is maintaining at 124.85 bar and the condensation pressure is decreased from 0.11 bar to 0.04 bar, the result is a decrease of 0.49 kg/s.

4. CONCLUSION

According to the energetic analysis performed in the regeneration cycles with six and seven heaters, the result is a difference in the values of SSC, SFC and UTC which is less than 0.5%, likewise, it shows an increased thermal efficiency when heaters are added to the operating line, for preheating. As the heaters are added, the difference in increasing the thermal efficiency is reduced, leading to suggestions that eight heaters would not be desirable since it would have increased less than 0.5%. Also, it is concluded that the increase or decrease of the condensing pressure and the inlet of the turbine plays a role on the energy cost for generating a kW-h.

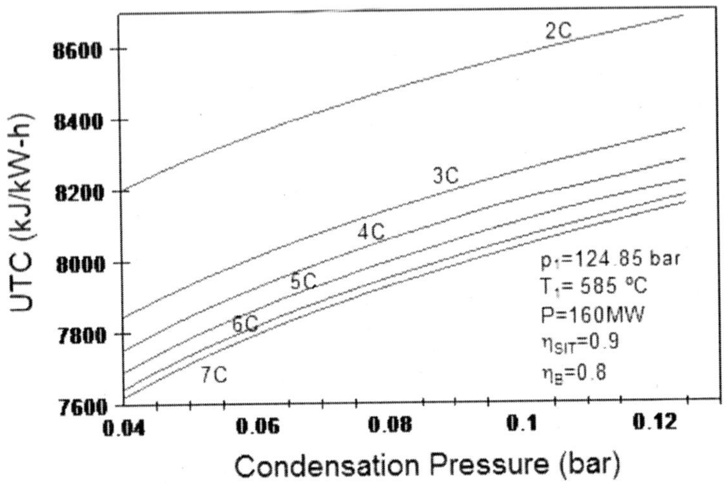

Figure 6. UTC to different pressure of condensation for steam cycles with regeneration.

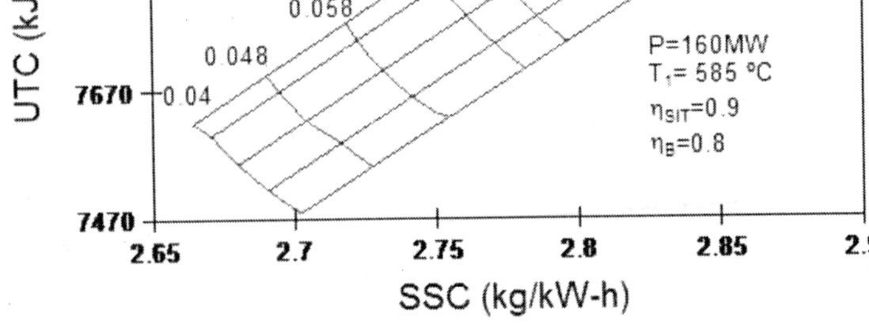

Figure 7. UTC and SSC to different condensation of pressure and different pressure in the inlet of the turbine of steam cycles with seven heaters.

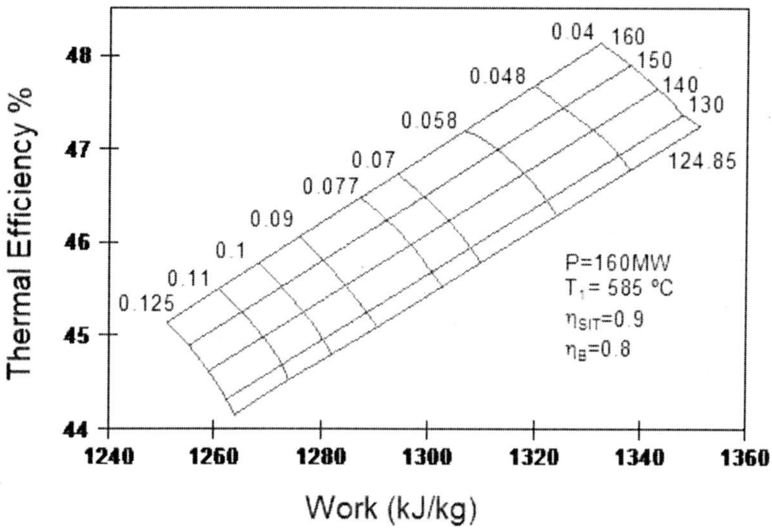

Figure 8. Thermal efficiency and useful work to different condensation pressures and turbine inlet steam cycle with seven heaters.

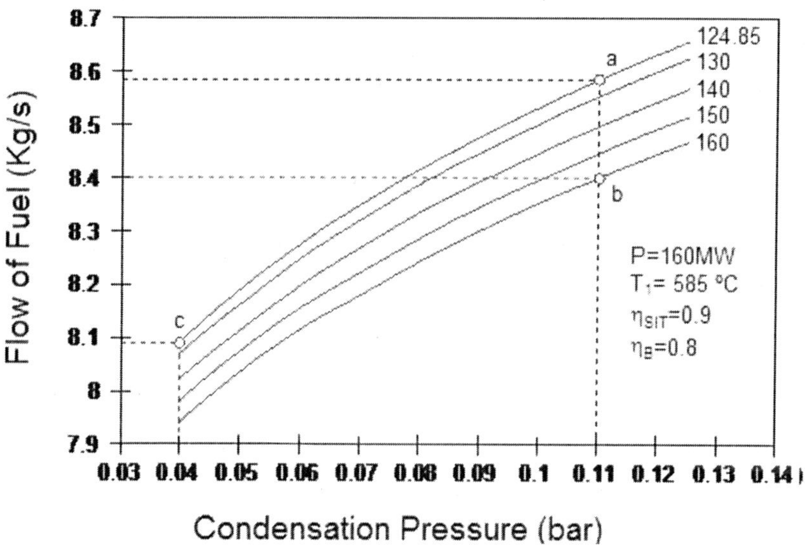

Figure 9. Fuel flow to different condensation pressures and turbine inlet steam cycle with seven heaters.

REFERENCES

1. Malik, A.S., Boyko, O., Atkar, N. and Young, W.F. (2001) A comparative study of MR imaging profile of titanium pedicle screws. Acta Radiologica, 42, 291-293.

2. Dincer, I. and Al-Muslim, H. (2001) Thermodynamic analysis of reheat cycle steam power plant. International Journal of Energy Research, 25, 727-739.

3. Federal Electric Commission (2012) Catalogue of the thermoelectric plant in Ciudad Laredo.http://www.sener.gob.mx/res/ PE_y_DT/pub/ 2012/PSE_2012_2026.pdf

4. Bohem, R.F., et al. (1992) Thermodynamics and design, analysis and improvement of energy systems. ASME AES, 27, 154-155.

5. Pereyra, M.S., Leyte, R.L., Aldaco, A.T. and González, R. (2005) Análisis energético y exergético del proceso de combustión en un generador de vapor. XI Congreso Internacional Anual de la SOMIM y IV Congreso Bolivariano de Ingeniería Mecánica, Morelia, 21-23 September 2005, 1-10.

6. Pereyra, M.S., Leyte, R.L., Aldaco, A.T., Velázquez, M.T. and Mata, J.M.Z. (2003) Análisis exergético de una central termoeléctrica. VIII Congreso Latinoamericano de Turbomáquinaria, Boca del Rio, Veracruz, 14 October 2003, 1-9.

Chapter 11

Environmental Aspects of Coal Combustion Residues from Thermal Power Plants

Gurdeep Singh[1]

[1] Department of Environmental Sciences and Engineering, Indian School of Mines, India

1. INTRODUCTION

Electricity is an essential need of any industrial society and no nation can progress without adequate supply of power. Growth in its demand during the past decades has been phenomenal and has outstripped all projections. In India also, there has been impressive increase in the power generation from a low capacity of 1330 MW in 1947 at Independence to about 81,000 MW at end of March 1995[1] and at about 2,05,340 MW at the end of June 2012 [2]. However, despite this substantial growth there still remains a wide gap between demand and supply of power which is expected to worsen in the years and decades to come. We already are experiencing shortage of nearly 8% of the average demand and 16.5% of peak demand. Moreover, with the quantum jump expected in demand for power in the future due to rapid industrialization and changing life styles of populace as a result of economic liberation, shortages shall further increase unless immediate steps are taken to increase power production.

1.1. The Energy Scene

Per capita consumption of energy in India is one of the lowest in the world. India consumed 570 kg of oil equivalent (kgoe) per person of primary energy in 2009-10 compared to 1090 kgoe in China and the world average of 1,688 kgoe [3]. India has declared to be moving towards energy security and independence in a few decades with ambitious energy generation targets. Total installed capacity source wise for June 2012, is give below in Table 1 and Figure 1 [2].

The total installed electric power generation capacity, public and private sector combined, is 2, 05,340.26 MW only. Of the total installed capacity, thermal power is highest at 116.333.38MW representing almost 57%, the remaining being either gas based, oil based, hydro or nuclear [4]. Coal has been identified the mainstay fuel for thermal power generation in India and will remain so for another two decades at least.

1.2. Coal Reserves and Demand Scenario

The Integrated Energy Policy Committee set up by GOI in 2004 emphasized the continuing dominance of coal in Indian Energy Security over the next 25 years [5]. India has vast reserves of coal and it is expected that coal remains a prime source of energy through the early part of 21st century The coal reserves in India up to the depth of 1200 meters have been estimated by the Geological Survey of India at 285.86 billion tonnes as on 1.4.2011 [6].. Only 12.5% of the resource is coking type and remaining 87.5% is non-coking or thermal coal. Coal deposits are mainly located in Jharkhand, West Bengal, Orissa, Chattrisgarh, Maharastra, Madhya Pradesh and Andhra Pradesh [6]. About 75% of the coal produced in the country is used for thermal power generation [7]. The coal demand is expected to rise to 2600 million tones at present level of consumption trend. The country's estimated demand of coal assessed by Planning Commission is 772.84 million tonnes for 2012-13 [6]. Majority of this demand will be met by our own resource of which more than 85% are high ash non-coking variety. Rapid increase in India's future power generation to meet the growing demand for domestic and industrial uses will be based on coal.

Table 1. Total Installed Capacity of Electricity Generation from Various Sources

Oxide	Class C (Wt%/std)	Class F (Wt%/std)
SiO_2	17.6 ± 2.7	52.5 ± 9.6
Al_2O_3	6.2 ± 1.1	22.8 ± 5.4
Fe_2O_3	25.2 ± 2.8	7.5 ± 4.3
CaO	"/>10	<10
MgO	1.7 ± 1.2	1.3 ± 0.7
Na_2O	0.6 ± 0.6	1.0 ± 1.0
K_2O	2.9 ± 1.8	1.3 ± 0.8
SO_3	2.9 ± 1.8	0.6 ± 0.5
LOI	0.06 ± 0.06	0.11 ± 0.14
Moisture	0.33 ± 0.35	2.6 ± 2.4

(Source: Scheetz et al., 1997)

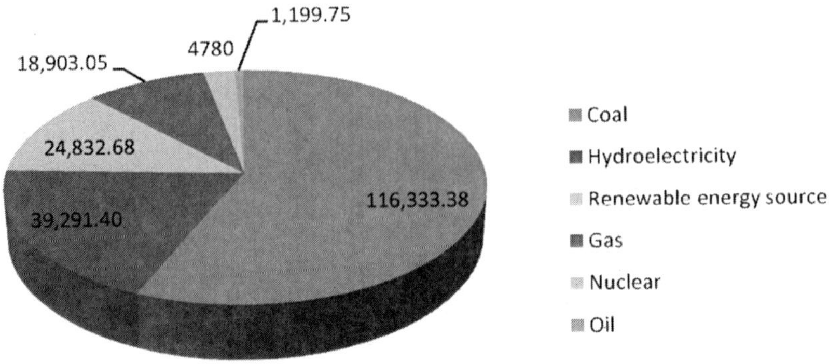

Figure 1. Total Installed Capacity of Electricity Generation from Various Sources (*Source: Central Electricity Authority, New Delhi, June, 2012)*

1.3. Coal Combustion Residues and Their Types

The process of coal combustion results in the generation of coal combustion residues (CCRs). India has diverse quality of coal reserves, which contain 30-55% ash. Annual production of CCRs has already crossed 120 MT and with the present rate of growth (8-10%) of power generation it is expected that this figure would be 175 MT by 2012 [8].

The quantum of CCRs generated in India is a matter of utmost concern. Out of the total CCRs generated, only 41% is being utilized [9]. Utilization that was around 1% in 1994 has shown a sharp increase in ten years period. However, the 41% utilization is still low as compared to more than 50% utilization in some other countries. One of the main reasons for the improper utilization of CCRs in India is its poor availability in usable forms and lack of characterization.

Coal combustion residues (CCRs) include materials that are left over after the burning of coal. They are fly ash, bottom ash, boiler slag and flue gas desulfurization (FGD) materials (wet or dry). The process of ash formation is influenced by the chemical composition of mineral matter in coal and its thermal properties. Other important factors include interaction between the inorganic and organic constituents of the particles and the physical and chemical environment through which these particles pass in the furnace. The distribution between the bottom ash and fly ash fraction is a function of the coal type, the boiler type and the type of boiler bottom.

1.3.1. Fly Ash

Fly ash is one of the incombustible mineral residues produced from the combustion of pulverized coal in the boilers. The particle size of the fly ash is varied between less than 1 to 100 μm. The fly ash particles are removed with the help of flue gases from the stack and generally captured by electrostatic precipitators and any other air pollution control equipments such as bag houses or wet scrubbers [10].

Classification of Fly Ash: According to ASTM C 618 (American Society for Testing and Materials), fly ash is classified into two categories such as Class C and Class F fly ash. Both the type of ash contains SiO_2, Al_2O_3 and Fe_2O_3. Class C fly ash contains greater than 50% and Class F fly ash contains approximately 70% of these oxides. The average composition of Class C and Class F fly ash *i s shown in Table 2.5* [11].

Class C Fly Ash: Class C fly ash is produced from the burning of sub bituminous coal or lignite. It has pozzolanic properties and self-cementing properties. Class C fly ash will harden and gain strength over time in the presence of water. Alkali and sulfate (SO_4) contents are generally higher in Class C fly ashes. It has higher lime (CaO) content (>10%) than Class F fly ash (<10%).

Class F Fly Ash: Class F fly ash is mainly produced by the anthracite and bituminous coal. It has pozzolanic properties and less than 10% lime content. Class F fly ash requires a cementing agent, such as hydrated lime or quicklime with the presence of water in order to produce cementitious materials.

1.3.2. Bottom Ash

Bottom ash is generally gray to blackish in colour. It is angular shaped and has a porous surface structure. Approximately 15-20% of total ash is bottom ash. It does not have any cementitious properties due to its larger size. Bottom ash consists of aggregated ash particles formed in pulverized coal boilers that are too large to be carried in the flue gases and fall through open grates to an ash hopper at the bottom of the boiler or impinge on the boiler walls [12]. Bottom ash is removed from the bottom of the boiler either in dry or wet state and is transported by the pipeline to handling system.

Table 2. Composition of Class C and Class F Fly Ash

S. No.	Percentage Utilization of Fly Ash	Target Date
1.	At least 50% of fly ash generation	One year from the date of issue of this notification
2.	At least 65% of fly ash generation	Two years from the date of issue of this notification.
3.	At least 85% of fly ash generation	Three years from the date of issue of this notification.
4.	100% fly ash generation	Four years from the date of issue of this notification.

(Source: Fly Ash utilization Amended Notification, MoEF, New Delhi, November 2009)

1.3.3. Pond Ash

After the generation of fly ash it is transported to the specific area termed as ash pond. Generally pond ash contains fly ash and bottom ash mixture in the ratio of 80:20. About 20-40 cubic meters water is require for the transportation of 1 tonne of ash. The coarser particles are generally settled down due to the action of gravity. Pond ash creates respiratory diseases and reduction of visibility during the time of summer. That's why water is spraying over the ash pond surface to control the air pollution.

1.3.4. Boiler Slag

Boiler slag is made up of black, hard and angular particles that have a smooth and glass like appearance. It is the molten ash collected at the bottom of the cyclone and slag tap that is mixed with water. When the molten slag comes in contact with the water, it fractures, crystallizes, and forms pellets. Boiler slag is mainly used for various applications such as fill material for structural applications, component of blasting grit and raw material in concrete products etc.

1.3.5. Fbc (Fluidized Bed Combustion) Ash

The FBC is a technology employed to reduce the amount of sulphur released into the atmosphere while burning sulphur-rich coal. FBC ash is formed when coal is burnt in the presence of crushed limestone as a bedding material in fluidized form. It consists of unburned coal, ash, and spent bed material. $SO_2$is converted to $CaSO_4$ through its reaction with the limestone during the time of combustion. Approximately lower temperature (815 - 870^{0}C) is require for fluidized bed furnace than the conventional coal fired furnaces (1400 -1600^{0}C). FBC technology has become extensively used for reducing SO_2 emissions from power plants due to its low cost [13].

1.3.6. Fgd (Flue Gas Desulfurization) Ash

FGD ash is the solid residue that results from a variety of processes used to control SO_2 emissions from boiler stacks. Apart from over 95% SO_2 removal capacity, this technology can also reduce the emission of other gases like hydrogen chloride and sulphur trioxide [14]. FGD ash generally produced in the scrubber by the reaction of limestone with SO_2 to form calcium sulfite. This calcium sulfite further oxidizes to form calcium sulfate (95% pure).Both of the compounds are produced in the scrubbers in wet form and then dried and processed for handling.

1.4. Implications Of Coal Ash Generation

Generally high ash coals are available for thermal power generation. Solid waste from combustion, mainly fly ash removed with electrostatic precipitators or bag houses and bottom ash collected in the boiler, have become an important environmental problem because of their high volume and physical and chemical characteristics. As per the available estimates the production of coal ash in India including both fly ash and bottom ash is about 120 million tones per annum which is likely to touch 200 million tons per annum by 2015 A.D [15]. Due to high ash content in Indian coal, emphasis is being laid now on setting up coal fired Super, Ultra and Mega Power Projects on pithead itself in order to minimize cost associated with bulk transport of ash laden coal [16]. The installed capacity is expected to increase to about 300,000 MW by 2017. Majority of these additions would be coal based which will further add to the burgeoning problems of ash disposal. A thermal power plant generating 1000 MW of electricity produces about 1.6 million tones of coal ash annually of which 80% is fly ash. The management of large volume of coal ash produced in power plants is a real challenge for the nation [17].

The power plants must either find purposeful utilization of these ashes or alternatively dispose them off site. Although coal ash possesses beneficial properties, both physical and chemical. Still serious concerns related to health, safety and environmental risks involving air and water quality prevail in the mind of mine planners, operators, regulators and environment groups [18]. Thus the utilization of coal ash has drawn considerable concern and attention of scientists, technologists, environmental groups, government, regulators etc. Till the early 1990's only a very small percentage (3%) of the fly ash was used productively in

India and the balance material was being dumped in slurry form in vast ash ponds close to power plants. The numbers of governmental and institutional actions taken since then have increased the ash utilization to 50% during 2010-2011 [19].

2. ENVIRONMENTAL REGULATIONS FOR THE FLY ASH UTILIZATION

To address the problem of pollution, caused by fly ash and to reduce the requirement of land for disposal of fly ash in slurry form in ash ponds, Ministry of Environment & Forests (MoEF) Government of India (GOI), has issued following notifications stipulating targets for utilization of the fly ash to achieve 100% utilization in phased manner [20].

2.1. Ministry Of Environment and Forests — Notification

Published in the Gazette of India, Extraordinary, Part II, Section 3, Subsection (ii), New Delhi, the 6th November, 2008.

S.O. 2623 (E)._Whereas by notification of the Government of India in the Ministry of Environment and Forests number S.O. 763(E), dated the 14th September, 1999 (hereinafter referred to as the said notification) issued under sub-section (1) and clause (v) of sub-section (2) of section 3 and section 5 of the Environment (Protection) Act, 1986 (29 of 1986), the Central Government, issued directions for restricting the excavation of top soil for manufacture of bricks and promoting the utilisation of fly ash in the manufacture of building materials and in construction activity within a specified radius of one hundred kilometres from coal or lignite based thermal power plants [21]; (2) All coal and, or lignite based thermal power stations and, or expansion units in operation before the date of this notification are to achieve the target of fly ash utilization as per the given below:

Table 3. Target of Fly Ash Utilization

S. No.	Percentage Utilization of Fly Ash	Target Date
1.	At least 50% of fly ash generation	One year from the date of issue of this notification
2.	At least 65% of fly ash generation	Two years from the date of issue of this notification.
3.	At least 85% of fly ash generation	Three years from the date of issue of this notification.
4.	100% fly ash generation	Four years from the date of issue of this notification.

(Source: Fly Ash utilization Amended Notification, MoEF, New Delhi, November 2009)

The unutilized fly ash in relation to the target during a year, if any, shall be utilized within next two years in addition to the targets stipulated for those years. The balance unutilized fly ash accumulated during first four years (the difference between the generation and the utilization target) shall be utilized progressively over next five years in addition to 100% utilization of current generation of fly ash.

(3) New coal and, or lignite based thermal power stations and, or expansion units commissioned after this notification to achieve the target of fly ash utilization as per given below:

The unutilized fly ash in relation to the target during a year, if any, shall be utilized within next two years in addition to the targets stipulated for these years. The unutilized fly ash accumulated during first three years (the difference between the generation and utilization target) shall be utilized progressively over next five years in addition to 100% utilization of current generation of fly ash.

MoEF's Notification of 3rd November, 2009 i.e. [22]

- All thermal power stations in operation on the date of notification should have achieved the target of 60% fly ash utilization within two years from the date of notification i.e. by 3rd November, 2011; and

- All new thermal power stations which have come into operation after the date of notification should have achieved the target of 50% of fly ash utilization within one year of their commissioning.

No person or agency shall within 50 kilometers (by road) from coal or lignite based thermal power plants, undertake or approve showing of mine without using at least 25% of fly ash on weight to weight basis, of the total stowing materials used and this shall be done under the guidance of the Director General of Mines Safety (DGMS) or Central Mine Planning and Design Institute Limited (CMPDIL).

Provided that such thermal power stations shall facilitate the availability of required quality and quantity of fly ash as may be decided by the expert committee referred in subparagraph (10) for this purpose.

8(ii) No person or agency shall within fifty kilometers (by road) from coal or lignite based thermal power plants, undertake or approve without using at least 20% of fly ash on volume to volume basis of the total materials used for external dump of overburden and same percentage in upper benches of back filling of opencast mines and this shall be done under the guidance of the Director General of Mines Safety (DGMS) or Central Mine Planning and Design Institute Limited (CMPDIL).

Table 4. Target of Fly Ash Utilization

S. No.	Fly ash utilization level	Target Date
1.	At least 50% of fly ash generation	One year from the date of commissioning.
2.	At least 75% of fly ash generation	Two years from the date of commissioning
3.	100% of fly ash generation	Three years from the date of commissioning.

(Source: Fly Ash utilization Amended Notification, MoEF, New Delhi, November 2009)

2.2. Strategies For Coal Ash Utilization

Central Electricity Authority (CEA) has been working out the straggles on behalf of Government of India for effective utilization of the coal ash [15]. A large number of technologies have been developed for gainful utilization and safe management of fly ash under the concerted efforts of Fly Ash Mission of the GOI since 1994. As a result, the utilization of fly ash has increased to over 73 million tonne in 2010-11. Fly ash was moved from "hazardous industrial waste" to "waste material" category during the year 2000 and during November 2009, it became a saleable commodity. Fly ash utilization has started gaining acceptance, it being 55.79% during 2010-11.

The areas of concern include:

- Improving the collection efficiency of the ESP & of quality of fly ash generated.

- Need for development and implementation of systems for collection of classified fly ash, Certification of its quality for value addition and bulk environment friendly transportation options.

- Need for development of schemes for collection of dry bottom ash and its effective utilization.

- Need to develop energy efficient ash slurry pumps capable of handling dense ash slurry.

- Open trucks are used to transport fly ash for manufacture of building products. There is a need to develop efficient bulk transportation options for supply of fly ash from power plant to the end user. For 100% fly ash utilization at the generating stations

- Technologies are to be developed for demonstration of bulk utilization options of fly ash in roads and embankments, mine fills, sea erosion tetrapods, roller compacted concrete.

- Guideline standards to be developed to ensure quality assurance in value added products from fly ash, viz., bricks, blocks, pavers, kerbstones, tiles, etc.

- Development and application of high value added utilization of fly ash such as:
    - Extraction of cenospheres (A cenosphere is a lightweight, inert, hollow sphere filled with inert air or gas, typically produced as a by-product of coal combustion at thermal power plants.)
    - Extraction of titanium oxide, Alumina
    - Development of composite materials, acid/fire resistant bricks /tiles
    - Development of abrasion resistant materials
    - Value added building materials
    - Agriculture amendments, etc.
    - –**"Incubation centers"** should be set up for technology validation.
    - –**"Self sustaining technology demonstration centers"** to be established for technology propagation schemes.
    - –Encourage **"Industry–Institute interactions"** for entrepreneur development, awareness, training programmes and workshops.
    - –Induction of **"fly ash subject in academic curriculum"** of Engineering, Architecture, and Post Graduate Science Courses.

2.3. Modes Of Ash Utilization During 2010–11

The major modes in which ash was utilized during the year 2010-11 along with utilization in each mode is presented in the pie diagram is given in Figure-2 below [15]:

It may be seen from above table that the maximum utilization of fly ash to the extent of 48.50% has been in Cement sector, followed by 12.73% in reclamation of low lying area, 11.65% in roads & embankments etc. The utilization of fly ash in mine filling was 8.26% and in making fly ash based building products like bricks, tiles etc was only 6.3%. These two areas have large potential of ash utilization which needs to be explored for increasing overall ash utilization in the country.

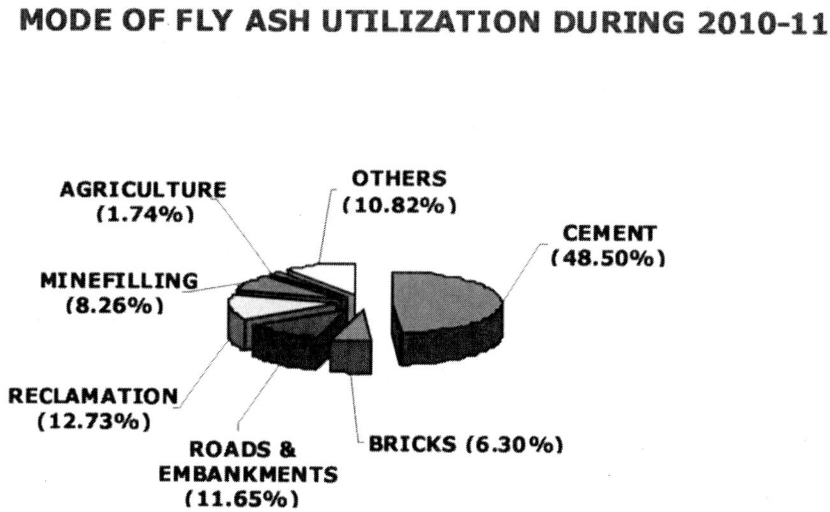

(Source: Central Electricity Authority, New Delhi, December, 2011)

Figure 2. Mode of Fly Ash Utilization during 2010-11.

2.4. Progressive Fly Ash Generation & Utilization During The Period From 1996–97 To 2010–11

The progressive ash generation at coal/lignite based thermal power stations and its utilization for the period from 1996-97 to 2010-11 as per data received in CEA from power utilities is given in Figure 3below [15]:

3. ENVIRONMENTAL ASPECTS OF TRACE ELEMENTS LEACHING

There has been widespread concern of ground and surface water contamination due to trace elements leaching in and around the ash ponds. Leaching aspects of trace elements are presented here which are required to screen the coal combustion residues (Fly as-, bottom- & pond – ash} on environmental angle for its utilization.

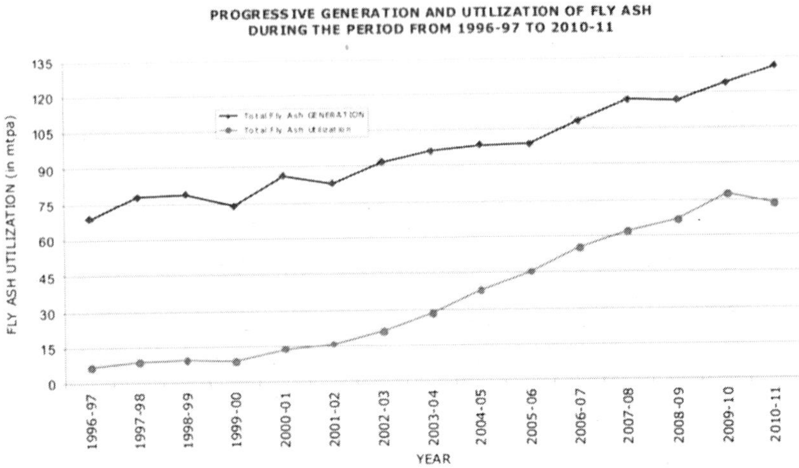

(Source: Central Electricity Authority, New Delhi, December, 2011)

Figure 3. Progressive Generation and Utilization of Fly Ash during the Period from 1996 – 97 to 2010 – 11.

3.1. Trace Elements In Fly Ash

Studies of trace elements and the elements present in fly ash are distributed into the fractions of the fly ashes based on volatilization temperature [23]. It is found that elements appear to partition into three main classes.

Elements that are not volatilized and reported equally in both fly ash and bottom ash. These elements include Al, Ba, Ca, Ce, Co, Cu, Fe, Hf, K, La, Mg, Mn, Rb, Se, Si, Sm, Sr, Th, Ta, and Ti.

Elements that are volatilized on combustion and preferentially get adsorbed on the fly ash as flue gas cools down. These include As, Cd, Ga, Mo, Pb, Sb, Se and Zn.

Elements that remain almost entirely in the volatilised state tend to escape to the atmosphere as vapours. These are Hg, Cl and Br.

Fly ash is an alumino silicate glass consisting of the oxides of Si, Al, Fe and Ca with minor amounts of Mg, Na, K, Zn and S and various trace elements. The concentration associated with the ash may be either adsorbed on the surface of particle or incorporated into matrix [24]. A mechanism that appears to be common for all ashes during their formation is the condensation of metal and metalloid vapours on refractory core materials. As the ash particles and gas stream exist from the combustion chamber and proceed upto the flue gas, this results in locally higher concentrations of many trace elements at the surface of ash particles and accounts for the generally higher concentration of these elements as particle size decreases [25]. The association between trace elements and major elements/minerals may be an important factor in determining the leachate composition of water in contact with ashes [26].

It is recognised that the health hazards and environmental impacts from coal fired thermal power stations result from the mobilization of toxic elements from ash [18, 24]. The large amount of ash that accumulates at thermal power plants, its possible reuse and the dispersion and mobilization of toxic elements from it, requires greater attention. Mobilization of various elements from the ash into the environment depends on climate, soils, indigenous vegetation and agricultural practices [27].

3.2. Leaching Aspects Of Fly Ash

The leaching characteristics of fly ash are mainly controlled by factors such as its chemical composition, mineralogy and morphology. Fly ash from thermal power plants vary in chemical composition not only from plant to plant but also within the same plant [26, 28]. Chemically fly ash consists of Si, Al, Mg, Ca, K, Ti and Fe in greater proportion with many trace elements as V, Mn, Cr, Cu, Ni, As, Pb, Cd and smaller quantity of various potential toxic elements. Chemical composition study shows mostly the presence of four major elements viz. aluminium, silicon, iron and calcium in the fly ash. Others such as potassium, magnesium, barium, cobalt, cadmium, zinc, molybdenum, lead etc. are present in traces [29]. Though in the traces, compared to the original coal, most of the elements are enriched in the fly ash, giving birth to the growing environmental concerns in the disposal and utilization environment due to release of trace/heavy elements metals [30].

Crystalline phases in fly ash as major constituent is quartz and others such as mullite, magnetite, hematite etc. are also present as minor constituents. The mineralogy of fly ash refers to both amorphous and crystalline phases and other mineral fractions [29]. The leaching behavior of fly ash can be related to their mineralogical characteristics Most of the toxic elements reside on major phases and are easily available during leaching [26].

Morphologically, particles of fly ash may be irregularly shaped particles, solid spheres and cenospheres. Morphology of fly ash similarly, controls the leaching of toxic elements/heavy metals. Morphology of fly ash is related to the conditions of combustion and composition of inorganic materials in coal. Particle morphology is one of the most important morphological aspects of fly ash. In this particle size particle distribution also holds its importance. Finer particles mean more surface area and so more concentration of condensing elements. In fact, specific surface area of fly ash is determined by both particle size and particle morphology.

The leaching of fly ash is a time dependent phenomenon. The initial leaching of the fly ash can be characterized by the surface hydrolysis and the dissolution of reactive phases formed under high temperature combustion. A close examination of the leaching studies show a rapid early dissolution followed by a later, slower release of the elements. The water-soluble fraction of a combustion residue may reflect the early dissolution process in the natural environment. In fact, the early dissolution mainly involves the soluble salts or the oxides on the particle surface of the fly ash. So, the dominant features of the initial dissolution stage are a high dissolution rate and the solution chemistry being controlled by buffering components of the fly ash. At this stage of high dissolution rate, release of salts or heavy metals associated with surface phases occur. The long term leaching of the fly ash occurs primarily in the alluminosilicate glass and some acid-soluble magnetic spinel phases. These phases usually constitute the matrix of the fly ash. The dominant leaching features of the matrix phases are a low dissolution rate and phase alteration over a long period of time [26].

Present study on environmental characterization is in continuation with earlier studies, to evaluate leaching of trace elements from coal ashes from a few Thermal Power Stations situated in eastern India are presented in this study [26, 28, 31]. The possible water contamination is also envisaged through the leachate analysis from ash pond disposal sites in real life situation.

3.3. Leaching Chemistry

Short-term leaching (shake) tests 24 and open column percolation leaching experiments were carried out on the fly ash samples to ascertain its leachate chemistry as briefly described below.

3.3.1. Strong Acid Digest Test

This is short term leaching study and is carried out in the presence of strong acids to provide the available concentration levels of trace/heavy elements in the samples.

For the purpose of the experiment 0.5gm fly ash sample is taken in a conical flask and to it is added 10ml nitric acid and 2ml perchloric acid. Mixture is then heated till dryness on hot plate. The conical flask is covered during the process of heating by funnel. The dried residue in the conical flask is then boiled with 2ml HNO_3 and then filtered. This is repeated with distilled water, warmed up and filtered till no residue is left in the conical flask. The filter paper in the silica dish is kept in the muffle furnace and heated to 850°C. Silica dish is then allowed to cool with residue in it and residue is collected for further analysis. The filtered solution is stored in polypropylene beaker for further analysis (elemental).

3.3.2. Astm Shake Test

This is the rigorous and short term leaching study. This test is run for twenty four hours and deionised water is used as the leaching medium. Shake tests can only be useful to a limited extent, and with variability in results in identifying the elements most likely to leach out of a material and to estimate the equilibrium constants for some of the reactions that takes place during the test. However, shake tests are poor indicators of the conditions that might be expected in the field [32].

In this test 80gm fly ash sample is taken and put in the measuring bottle and to it is added two litre of distilled water. The bottle with the sample is then taken in the rotary agitator and rotated for twenty-four hours and this way sample with water is agitated for thorough mixing to result leaching. The extraction is performed in triplicate on each fly ash sample and the three replicates are then mixed to get the composite leachate sample. The leachate so obtained is filtered using Whatmann No. 42 filter paper. The filtrate so collected is then stored in the polypropylene bottle for further analysis. Once the potentiometric analysis is over, few drops of 6N nitric acid are added to the leachate collected in polypropylene bottle to avoid contamination and the bottle stored to be used for further elemental analysis of the leachate by AAS.

3.3.3. 30–Day Shake Test

This test is similar to 24-hr shake test and is carried over a period of 30 days and is intended to indicate the solubility of elements that may reach equilibrium with the surrounding water more easily. This test is required by some regulatory agencies of some of the States in the USA.

3.3.4. Toxicity Characteristics Leachate Procedure

The toxicity characteristic leachate procedure (TCLP) requires the use of an extraction fluid made of buffered acidic medium to run the test. For this the selection of the extraction fluid is made prior to conducting the test. Once the extraction fluid to be used in the test is determined, 40g fly ash sample is taken and then extraction fluid equal to twenty times the amount of sample taken is added in the zero head extractor under pressure. The system is tightly closed and then placed in an end-over-end rotary shaker for 18 hours, rotating at 30 ±2 rpm at a room temperature of about 25°C. Leachate after the said period of shaking is pressure filtered using 0.7 micron pore size filter paper.

3.3.5. Modified Synthetic Leachate Procedure (Slp)

This test is a modified SLP rather than a standard SLP. Here, unlike the standard SLP test which make use of a mineralized synthetic leaching medium prepared from deionized water, water from an actual field site (containing ions and impurities similar to those found in groundwater in an area of interest) is used. This test aims to detect the ion exchange reactions that can only be observed in the field like ion bearing water. This test is run for 24 hours.

3.3.6. Astm Column

In the ASTM column procedure, one pore volume of distilled water is forced through a packed column of fly ash each day in a saturated upflow mode. Leaching in this column is conducted under a nitrogen atmosphere and thus present leaching in an oxygen poor environment. In the field, this type of leaching would occur below the water table where there is low concentration of oxygen. One pore volume corresponds to the void space between the material grains in the test column. The rate of percolation of the leaching medium regardless of the hydraulic conductivity of the material is controlled by applying variable nitrogen pressure. The test is run for 16 days and leachate collected after 1, 2, 4, 8, 16 days of leaching [33].

3.3.7. Open Percolation Column Experiments

In these experiments, deionized water is percolated through a packed column of fly ash in the presence of oxygen at a rate which depends on the natural permeability of the material. The open columns for leaching experiments are made of PVC pipe four inches in diameter and two feet in length. The column setup involved packing the coal ash material at optimum moisture and density conditions as determined by the Proctor test. The fly ash material is packed into the column in two inch lifts with a 2" x 2" wooden rod, about 4 feet long. Each packed layer is scarified, by lightly scraping the top of the packed layer with a long thin rod to ensure proper interlocking of the material. The top six inches of the column was left unpacked to allow for the addition and maintenance of the leaching medium. About 200 ml of leaching medium (de-ionized water) is added at the top of the column once every alternate day to maintain sufficient supply of water to the packed coal ash material. The top end of the column is exposed to the atmosphere and the bottom end is connected to quarter inch tubing. The columns discharged the leachates through this tubing into the 250 ml polypropylene beakers. The leachates are collected in these beakers and analyzed.

3.4. Elemental Analysis Of Leachates

The leachate samples are filtered and acidified with 2 ml of nitric acid and then preserved in polypropylene sampling bottles. The samples are kept in a refrigerator until further analysis. Sodium and potassium were determined using flame photometer. Concentration levels of trace elements were evaluated using Atomic Absorption Spectrophotometer (AAS). Working/standards solutions were prepared according to instructions given in the operation manual of the GBC-902 AAS [34]. Optimized operating conditions such as lamp current, wave length, slit width, sensitivity, flame type etc. as specified in the manual, are used for analysis of a particular element. -AAS standards are used for standardization and calibration of AAS. Three standards and a blank of the concerned element are used to cover the range 0.1-0.8 Abs. The calibration is performed by using the blank solution to zero the instrument. The standards are then analyzed with the lowest concentrations first and the blank is run between standards to ensure that baseline (zero point) has not changed. Samples are then an analyzed and their absorbance recorded. The calibration is performed in the concentration mode in which the concentration of sample is recorded. ICP-MS can be used to arrive at the precise and quick results wherever affordable.

4. LEACHATE ANALYSIS RESULTS

Comparative evaluation of short-term leaching (shake) tests is presented in Table 2. Acid digest data provides the available concentration levels of Trace elements in fly ash whereas in comparison to this shake tests resulted significantly lower concentration levels particularly in 24 hr. ASTM, 30 Day and SLP shake tests. TCLP leachate data however, gives rise to significantly higher levels as this involved leaching in slightly acidic buffered conditions and as such does not reflect the actual behavior. It may be emphasized that these shake test presented the accelerated leaching because of 1:20 solid to liquid (leachant) ratio. Nevertheless these short term (shake) leaching tests provide an immediate and rapid indication of leachable concentration levels of trace elements from fly ash.

Analysis of twenty two elements were carried out from each of the leachate samples collected from open column experiments and the observations are summarized in Tables 3 & 4 for fly ash, pond ash and actual ash pond leachates, respectively. It is noticed from the observations that the concentration of thirteen elements, namely, chromium, nickel, cobalt, cadmium, selenium, aluminum, silver, arsenic, boron, barium, vanadium, antimony and molybdenum were below the detection limit (.001 mg/l) in the entire study period. Among the other nine elements only calcium and magnesium were observed in the leachates throughout the study period while the concentration of other elements showed a decreasing trend to below detection limit (.001 mg/l). In the leachates from actual ash ponds, lead and manganese were found absent but iron, calcium, magnesium, sodium, potassium, copper and zinc were present throughout the study period.

A comparison of the concentration levels observed in the leachates of fly ash, pond ash and also leachates from actual ash pond disposal site with the permissible limits as per IS:2490, is presented inTables 3 & 4 which indicates that the concentration levels of all the elements during the entire study period were either below detection limits (BDL) or below the permissible limits.

It can be inferred that no significant leaching occurs and toxicity is manageable with respect to trace elements both in the ash pond disposal site as well as in the open column leaching experiments. Further, analysis results of leachates from open column percolation experiments resemble closely with those of actual ash pond leachates. The physical set up of the open columns more closely resembles with because the flow of the leaching medium is influenced by gravity alone and the solid to liquid ratio is more close to the field situation. Hence, open column leaching experiments may be used in predicting the long term leaching behaviour that can be observed in the field. Fly ash leachates as generated from open percolation column leaching experiments and those from ash pond disposal site closely resemble and as such do not pose any significant environmental impacts in the disposal system. Overall, fly ash would not seem to pose any environmental problem during its utilization and/or disposal. Leaching pattern trace elements over three years open percolation column experiments is depicted in Figures 2-13 – which clearly reflect that trace elements leaching is not a significant concern and coal combustion residues can be appropriately utilized as these are established generally as environmentally benign material.

5. CONCLUDING REMARKS

On the basis of the study of the leaching of trace elements from coal ashes, following conclusions can be drawn:

1) In the study period of about three (3) years there was practically no leaching of thirteen elements namely, chromium, nickel, cobalt, cadmium, selenium, aluminum, silver, arsenic, boron, barium, vanadium, antimony and molybdenum from all the ash samples.

2) Out of the nine elements found in the leachates only calcium and magnesium were found to be leaching in the entire period. The leaching of other seven elements namely, iron, lead, copper, zinc, manganese, sodium and potassium was intermittent. The leaching of sodium and potassium practically stopped due to first flash phenomenon after 35 and 40 days, respectively. It is emphasised long-term leaching results should be considered to arrive at the environmental screening of such materials.

3) The concentration of the elements in the leachates was invariably well below the permissible limits for discharge of effluents as per IS: 2490 and also for drinking water as per IS: 10500.

Overall, the fly ash samples from various Thermal Power Stations evaluated in this study were found to be environmentally benign and can be engineered for their bulk utilization particularly for mined out areas reclamation and for soil amendment for good vegetation.

The Centre of Mining Environment at ISM Dhanbad is currently engaged in evolving low technology high volume field demonstration to show that fly ash particularly fly ash can be disposed and utilized as fill material in an environmentally acceptable way in reclamation of abandoned mines [35, 36]. Fly ash has been successfully used as backfill in material in reclamation of mined out (goaf) area and at the top of the surface Helipad is set up, at Jamadoba Tata Steel Mining Area [37]. This has attracted a lots of public attention as a result of aesthetic and scenic value provided in the Eco-Park thus resulted. Similarly a considerable quantity of fly ash has also been utilised at Ghantotand OB Dumps and Damoda worked out opencast mine sites. At these sites trace elements leaching even after three years of monitoring does not seem to pose any environmental problem. With these encouraging results cooperative arrangements are being made by power utilities and mining authorities for utilisation of fly ash in reclamation of various mined out areas in SECL, MCL, NCL, BCCL, SCCL etc. [38].

Table 5. Comparative Leachate Analysis Results of Shake Tests for Fly Ash

Parameter	Acid digest	TCLP	24-hr	30-D	SLP	SLP Blank
pH	--	4.29	6.22	6.26	7.08	7.06
Conductivity	--	3.56	0.096	0.099	0.075	0.085
TDS	--	1.78	48	63	51	58
Iron	82.41	0.089	0.045	0.05	0.029	0.038
Lead	BDL	BDL	BDL	BDL	BDL	BDL
Magnesium	7.579	7.512	3.15	4.50	2.4	2.8
Calcium	304.00	304.13	3.37	5.12	63	70
Copper	0.094	0.215	BDL	BDL	BDL	BDL
Zinc	0.276	2.140	0.020	0.025	0.180	0.185
Manganese	0.638	0.314	0.031	0.030	0.021	0.028
Sodium	54.60	1452	39.80	41.10	6	10
Potassium	7.60	8.10	2.80	5.36	3	4
Chromium	0.860	0.803	BDL	BDL	BDL	BDL
Nickel	0.118	0.112	BDL	BDL	BDL	BDL
Cobalt	BDL	BDL	BDL	BDL	BDL	BDL
Cadmium	BDL	BDL	BDL	BDL	BDL	BDL
Selenium	BDL	BDL	BDL	BDL	BDL	BDL
Aluminium	BDL	BDL	BDL	BDL	BDL	BDL
Silver	BDL	BDL	BDL	BDL	BDL	BDL
Arsenic	BDL	BDL	BDL	BDL	BDL	BDL
Boron	BDL	BDL	BDL	BDL	BDL	BDL
Barium	BDL	BDL	BDL	BDL	BDL	BDL
Vanadium	BDL	BDL	BDL	BDL	BDL	BDL
Antimony	BDL	BDL	BDL	BDL	BDL	BDL
Molybdenum	BDL	BDL	BDL	BDL	BDL	BDL
Mercury	BDL	BDL	BDL	BDL	BDL	BDL

BDL- Below Detectable Limit; Concentration of Elements in ppm; TDS in ppm; Conductivity in mmhos/cm

Use of fly ash as backfill material for reclamation of mined out sites provide benefits such as easy availability, cheaper to transport because empty coal carriers returning from the power plant can "back haul" it to the mine site.

From the standpoint of the power plant, this is essentially a waste material which requires large costs of handling and a disposal to comply with environmental regulations. From the environmental point, this waste material will go back to the same place where it was mined and use of this material serves as extra benefit to power plants. Studies are also in progress to use fly ash for agriculture development.

Table 6. Summary of the Leachate Analysis of Fly Ash from Thermal Power Stations# 1

Parameter	Open Percolation Column Experiments Leachates			Ash Pond Leachate	(IS: 2490, 1981)
	Samples				Inland Surface Water
	FA#A	FA#B	PA		
pH	5.97-10.51	5.82-9.10	5.86-9.03	6.95-8.26	5.5-9.0
Conductivity	0.042-0.750	0.037-0.820	0.052-0.920	543-796	-
TDS	21-375	19-410	30-460	272-400	2100
Iron	BDL-0.740	BDL-1.220	BDL-1.369	0.89-1.983	-
Lead	BDL-0.420	BDL-0.396	BDL-0.490	0.121-0.462	0.1
Magnesium	BDL-15.53	0.039-38.00	0.065-44.00	17-29	-
Calcium	1.00-103.92	0.265-189.20	0.798-102-20	21-58	-
Copper	BDL-0.190	BDL-0.068	BDL-0.090	0.023-0.055	3
Zinc	BDL-0.380	BDL-0.372	BDL-1.529	0.295-1.763	5
Manganese	0.009-0.057	0.010-0.105	0.007-0.076	0.027-0.089	-
Sodium	3-56	3-49	3-47	19-43	-
Potassium	2-42	2-36	2-33	7-51	-
Chromium	BDL	BDL	BDL	BDL	2
Nickel	BDL	BDL	BDL	BDL	3
Cobalt	BDL	BDL	BDL	BDL	-
Cadmium	BDL	BDL	BDL	BDL	2
Selenium	BDL	BDL	BDL	BDL	0.05
Aluminium	BDL	BDL	BDL	BDL	-
Silver	BDL	BDL	BDL	BDL	-
Arsenic	BDL	BDL	BDL	BDL	0.2
Boron	BDL	BDL	BDL	BDL	2
Barium	BDL	BDL	BDL	BDL	-
Vanadium	BDL	BDL	BDL	BDL	-
Antimony	BDL	BDL	BDL	BDL	-
Molybdenum	BDL	BDL	BDL	BDL	-
Mercury	BDL	BDL	BDL	BDL	0.01

BDL- Below Detectable Limit; Concentration of Elements in ppm; TDS in ppm; Conductivity in mmhos/cm

Table 7. Summary of the Leachate Analysis of Fly Ash from Thermal Power Stations# 2

Parameter	Open Percolation Column Experiments Leachates			Ash Pond Leachates	(IS: 2490, 1981)
	Samples				
	FA#1	FA#2	PA		Inland Surface Water
pH	4.98-9.92	4.38-8.90	4.71-8.92	7.2-8.58	5.5-9.0
Conductivity	0.060-0.962	0.070-0.848	0.036-0.973	645-892	-
TDS	30-481	35-424	30-487	320-445	2100
Iron	BDL-3.850	BDL-3.120	BDL-3.120	1.02-2.941	-
Lead	BDL-0.098	BDL-0.080	BDL-0.249	-	0.1
Magnesium	BDL-37.9	BDL-36.4	BDL-21.0	10-19	-
Calcium	1-87.6	2-72.2	2.12-48.0	18-46	-
Copper	BDL-0.094	BDL-0.088	BDL-0.052	0.011-0.047	3
Zinc	BDL-1.082	BDL-1.100	BDL-1.290	0.93-1.015	5
Manganese	BDL-0.099	BDL-0.092	BDL-0.069	-	-
Sodium	BDL-48	BDL-23	BDL-82	5-10	-
Potassium	BDL28	BDL-36	BDL-18	8-18	-
Chromium	BDL	BDL	BDL	BDL	2
Nickel	BDL	BDL	BDL	BDL	3
Cobalt	BDL	BDL	BDL	BDL	-
Cadmium	BDL	BDL	BDL	BDL	2
Selenium	BDL	BDL	BDL	BDL	0.05
Aluminium	BDL	BDL	BDL	BDL	-
Silver	BDL	BDL	BDL	BDL	-
Arsenic	BDL	BDL	BDL	BDL	0.2
Boron	BDL	BDL	BDL	BDL	2
Barium	BDL	BDL	BDL	BDL	-
Vanadium	BDL	BDL	BDL	BDL	-
Antimony	BDL	BDL	BDL	BDL	-
Molybdenum	BDL	BDL	BDL	BDL	-
Mercury	BDL	BDL	BDL	BDL	0.01

BDL- Below Detectable Limit,

Concentration of Elements in ppm,

TDS in ppm,

Conductivity in mmhos/cm

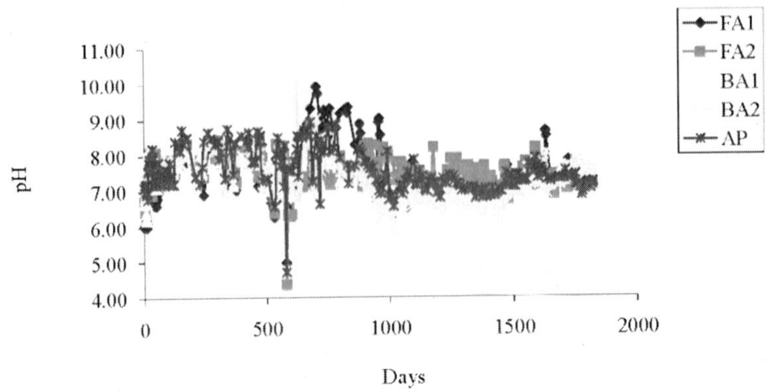

Figure 4. Open Column Leachate Analysis for pH

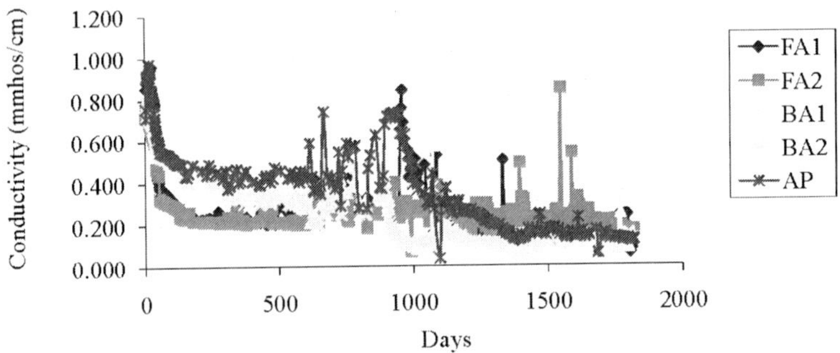

Figure 5. Open Column Leachate Analysis for Conductivity

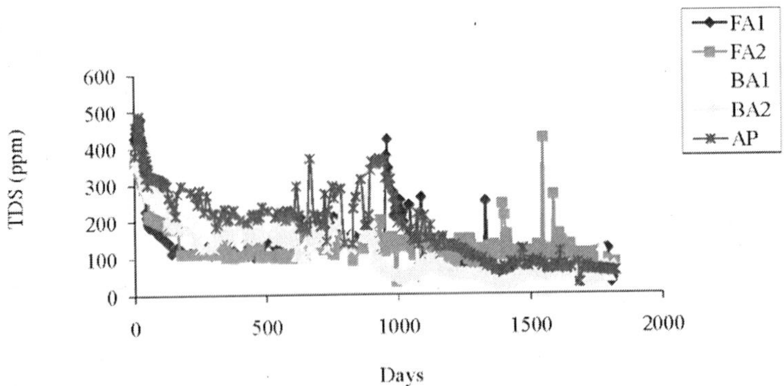

Figure 6. Open Column Leachate Analysis for TDS

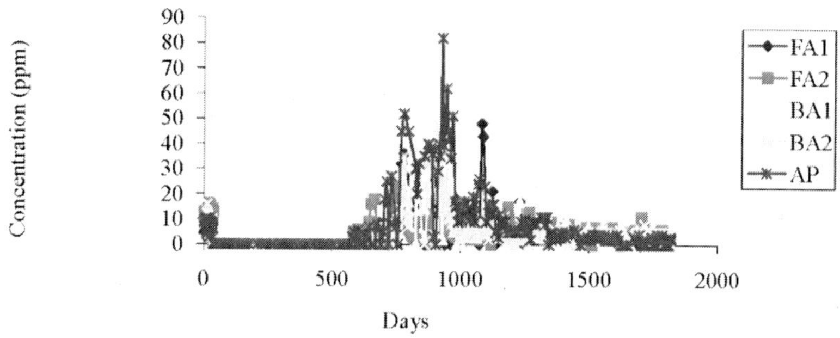

Figure 7. Open Column Leachate Analysis for Sodium

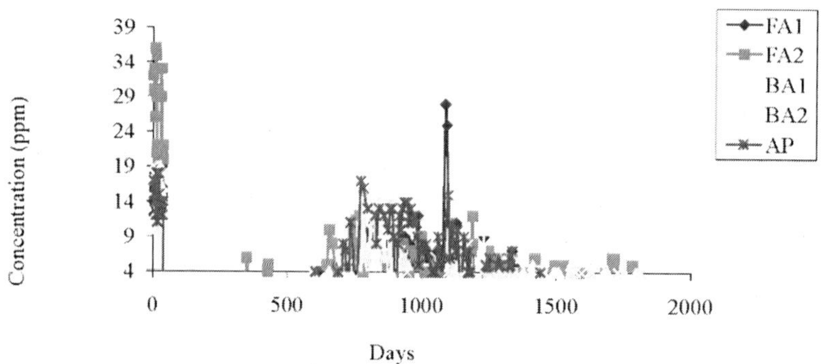

Figure 8. Opern Column Leachate Analysis for Potassium

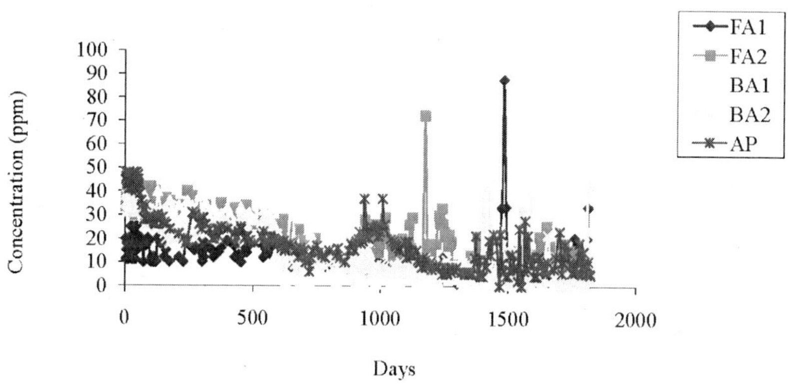

Figure 9. Open Column Leachate Analysis for Calcium

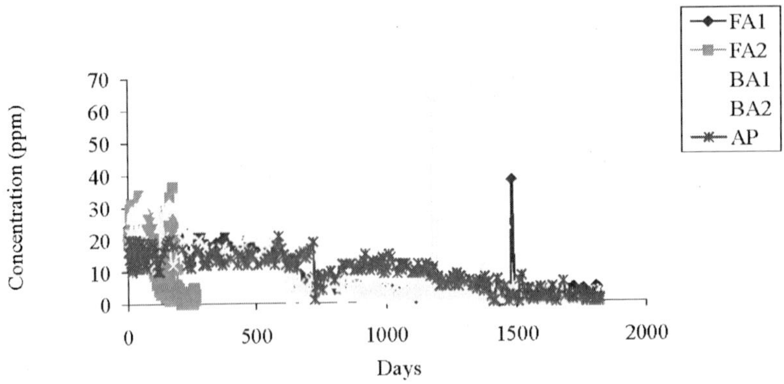

Figure 10. Open Column Leachate Analysis for Magnesium

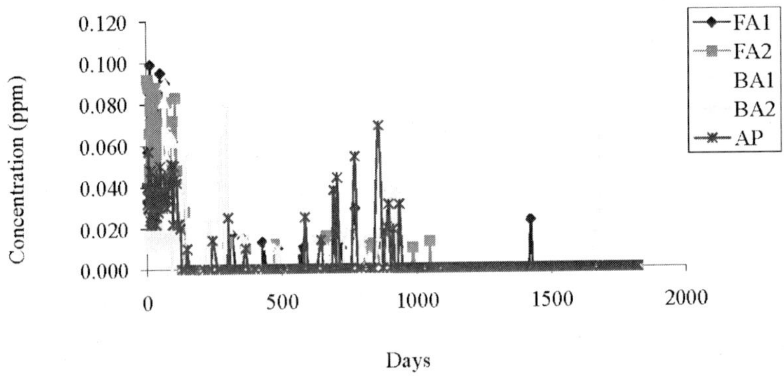

Figure 11. Open Column Leachate Analysis for Manganese

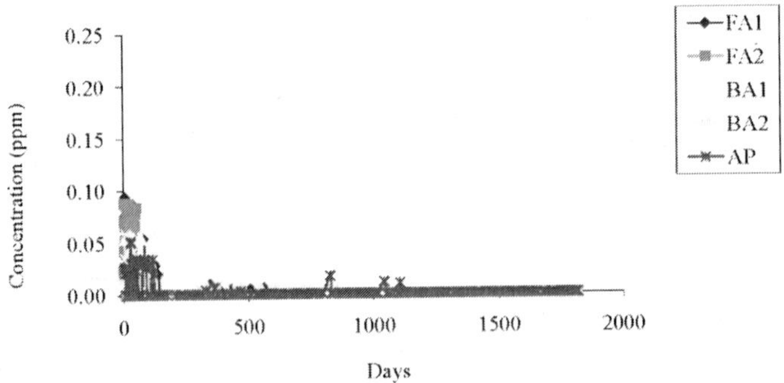

Figure 12. Open Column Leachate Analysis for Copper

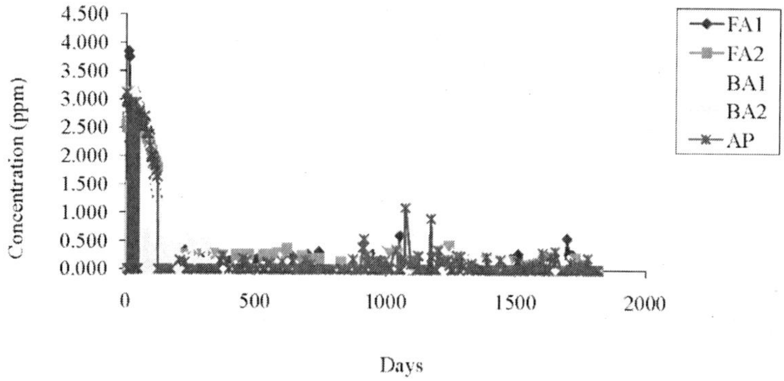

Figure 13. Open Column Leachate Analysis for Iron

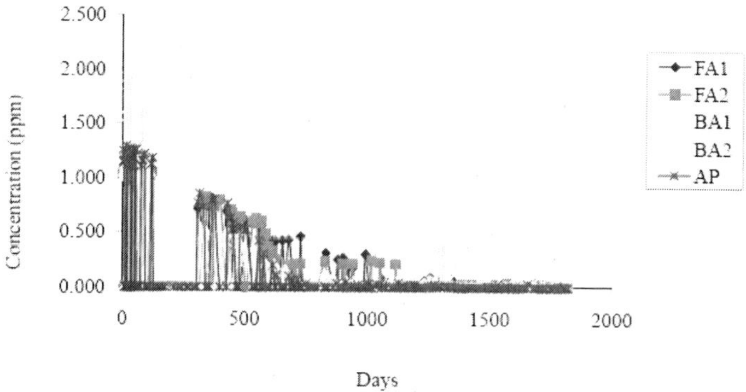

Figure 14. Open Column Leachate Analysis for Zinc

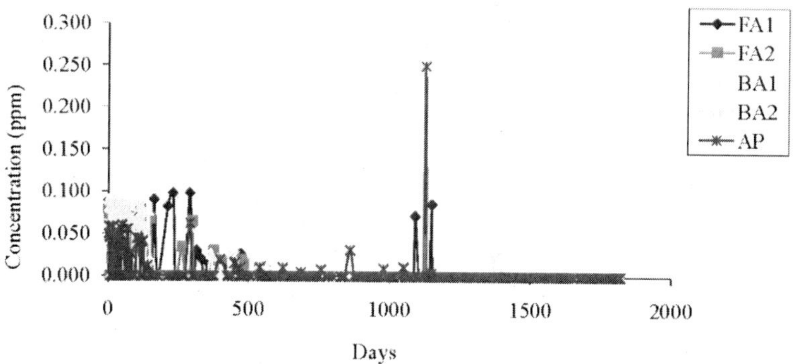

Figure 15. Open Column Leachate Analysis for Lead

REFERENCES

1. India Energy ForumPower India Year Book-2005 06

2. Ministry of Power http://www.cea.nic.in

3. Ministry of Petroleum and Natural GasAnnual Report, 2005

4. A Trehan, R Krishnamurthy, A Kumar, NTPC'S experience in ash utilization" Seminar on Fly Ash Utilization, New Delhi, 26 27March, 1996

5. Integrated Energy PolicyReport of The Expert Committee, Government of India Planning Commission, New Delhi, August 2006 1 137

6. Ministry of CoalAnnual Report, 2011 12

7. Ministry of CoalAnnual Report, 2005

8. www.tifac.org.in

9. V Kumar, and M Mathur, 2004Fly ash- gaining acceptance as building, proceedings of the Seminar on recent trends in Building materials, Bhopal, India, 55 65

10. C Heidrich, 2003Ash utilization-an Australian perspective. Proc. Of International Ash Utilization Symposium, Centre for Applied Energy Research, University of Kentucky.

11. ScheetzBarry, E., Michael J. Menghini, Rodger J. Hornberger, Thomas D. Owens and Joseph Schueck, 1997Utilization of fly ash. Proc. of Air and Waste Management Association. Toronto, Ontario, Canada.

12. http://wwwadaa.asn.au

13. H. L Wang, N. S Bolan, M J Hedley, and D. J Home, 2006Potential uses of fluidized bed boiler ash (FBA) as a liming material, soil conditioner and sulfur fertilizer. In Coal Combustion Byproducts and Environmental Issues. Eds. Sajwan, K. S., Twardowska, I., Punshon, T. and Alva A. K. Springer, New York, USA. 202 215

14. R. K Srivastava, 2000 S. O Controlling, emissions-a review of technologies. Prepared for the U.S. Environmental Protection Agency, EPA/600/R-00/093.

15. Central Electricity Authority (CEA)Report on Fly Ash Generation at Coal/Lignite Based Thermal Power Stations and its Utilization in the Country for the Year 2010-11" New Delhi, 1 18December, 2011

16. A Palit, R Gopal, S. K Dubey, and P. K Mondal, Characterization and utilization of coal ash in the context of Super Thermal Power Stations. Proceedings International Conference on Environmental Impact of Coal Utilization, IIT, Bombay, 154 155 1991

17. Patitapaban SahuCharacterization of Coal Combustion By-Products (CCBS) for their Effective Management and Utilization, Thesis, 2010

18. C. L Carloon, and D. C Adriano, Environmental Impacts of Coal Combustion Residues, Journal of Environmental quality 22 227 247 1993

19. Kumari VibhaEnvironmental Assessment of Fly Ash in its Disposal Environment at F.C.I, Sindri, Jharkhand (India)", M. Tech Thesis, Submitted to Indian School of Mines Dhanbad, 1998

20. http://envfor.nic.in/

21. Draft Notification on Fly ashS.O. 2623 (E), Published in the Gazette of India, Extraordinary, Part II, Section 3, Subsection (ii), New Delhi, dated 06 November, 2008

22. Fly Ash utilization Amended notification SO. 2804 (E), MOEF, New Delhi, dated 3rd November 2009

23. B. F Bahor, and H. J Gluskoter, 1973Boron in illite as an indicator of paleosalinity of Illinois coals, JSP, 43 945 956

24. D. F. S Natusch, J. R Wallace, and C. A Evans, Toxic trace elements: preferential concentration in Respirable particles". Science, 183 4121 203 204 1974

25. G. R Markowski, and R Filbly, Trace element concentration as a function of particle size in fly ash from a pulverised coal utility boiler" Environmental Science & Technology, 19 796 800 1985

26. Gurdeep Singh and Kumar SanjayEnvironmental evaluation on leaching of trace elements from coal ashes: a case study of Chandrapura Thermal Power Station". Journal of Environmental Studies and Policy 2 2000 2 135 142

27. Gurdeep Singh and Sanjay Kumar GambhirEnvironmental evaluation of fly ash in its disposal" environment. Proceedings, International Symposium on Coal- Science Technology, Industry Business & Environment, Nov. 18-19, 1996. editors. Kotur S. Narsimhan & Samir Sen. Allied Publishers Ltd. New Delhi. 547 556 1996

28. Gurdeep Singh and Kumar SanjayEnvironmental evaluation of coal ash from Chandrapura Thermal Power Station of Damodar Valley Corp." Indian J. Environmental Protection. 18 1999 12 884 888

29. A. L Page, A. A Elseewi, and I. R Straughan, Physical and chemical properties of fly ash from coal fired power plants, Res. Review, 71 83 120 1979

30. Gurdeep Singh and Kumari Vibha 1999 Environmental assessment of fly ash in its disposal environment at FCI, Ltd. Sindri, Poll. Res. 18 3 339 343

31. R. K Jain, Environmental Assessment of Coal Combustion By-products of Burnpur Thermal Power Station, M. Tech. Thesis Submitted to Indian School of Mines, Dhanbad, 1998

32. Annual Book of ASTM Standards 11 1990

33. Annual Book of ASTM Standards 04 1991

34. Atomic Absorption Spectrophotometer Operating ManualAAS, GBC-902, Australia. 1990

35. C Bradley, Paul, and Gurdeep Singh, Environmental evaluation of the feasibility of disposal and utilization of coal combustion residues in abandoned mine sites. Proceedings, First World Mining Environment Congress. Dec. 11-14, 1995. N. Delhi. Oxford & IBH Publ. Co. Pvt. Ltd. New Delhi. 1015 1030 1995

36. C Bradley, Paul, Gurdeep Singh, Steven Esling, Chaturvedula and Pandal, H. The impact of scrubber sludge on ground water at an

abandoned mine site. Environmental Monitoring and Assessment, 50 1998 1 13

37. A. P Singh, Environmental Evaluation of Coal Combustion Residues of Jamadoba Fluidised Bed Combustion Plant of TISCO with Special Emphasis on Stabilisation of Soil, M. Tech. thesis submitted to Indian School of Mines, Dhanbad, 2000

38. C Bradley, Paul, Gurdeep Singh and Chaturvedula, S. 1995. "Use of FGD by products to control subsidence from underground mines: groundwater impacts." Proceedings International Ash Utilization Symposium Lexington, Ky, USA, October 23 25 1995

Index